# Introduction to Environmental Engineering

# Introduction to Environmental Engineering

Edited by
**Bryson Walsh**

Larsen & Keller
www.larsen-keller.com

Introduction to Environmental Engineering
Edited by Bryson Walsh
ISBN: 978-1-63549-111-1 (Hardback)



Larsen & Keller
Published by Larsen and Keller Education,
5 Penn Plaza,
19th Floor,
New York, NY 10001, USA

**Cataloging-in-Publication Data**

Introduction to environmental engineering / edited by Bryson Walsh.
p. cm.
Includes bibliographical references and index.
ISBN 978-1-63549-111-1
1. Environmental engineering. 2. Green technology. 3. Environmental management.
I. Walsh, Bryson.
TA170 .I58 2017
628--dc23



The publisher's policy is to use permanent paper from mills that operate a sustainable forestry policy. Furthermore, the publisher ensures that the text paper and cover boards used have met acceptable environmental accreditation standards.

Printed and bound in China.

For more information regarding Larsen and Keller Education and its products, please visit the publisher's website www.larsen-keller.com

# Table of Contents

# Preface

This book elucidates the concepts and innovative models around prospective developments with respect to environmental engineering. It provides detailed information about the fundamentals of this subject. Environmental engineering refers to the science and methodology followed for protecting humans from harmful environmental components and also for protecting the environment from adverse effects of human activities and natural occurrences, by using engineering and scientific laws. Most of the topics introduced in this textbook cover new techniques and the applications of environmental engineering. It is a complete source of knowledge on the present status of this important field. This book, with its detailed analyses and data, will prove immensely beneficial to graduates and post-graduates involved in this area at various levels.

To facilitate a deeper understanding of the contents of this book a short introduction of every chapter is written below:

Chapter 1- Environmental engineering is a division of engineering that deals with the application of the principles of engineering for the protection of the environment as well as human populations from detrimental human and natural activities. It takes into account the impact that technological advancement has on the quality of the environment. This chapter introduces the reader to these themes and provides an overview of the discipline of environmental engineering.

Chapter 2- This chapter expounds on the major areas of environmental engineering including vehicle emissions control, ecological engineering, sanitation, industrial waste treatment and public health. These areas are important as they form the crux of the issues that are tackled by environmental engineering. The topics discussed in the chapter are of great importance to broaden the existing knowledge on environmental engineering.

Chapter 3- There are several issues to be considered by environmental engineers, some of which are resource depletion, climate change, global warming, environmental degradation, waste management etc. This chapter presents all these concerns and educates the reader about the local and global scenarios that impact environmental engineering.

Chapter 4- Environmental engineering uses several technologies to better the environment, reduce pollution, monitor natural ecosystems and mitigate the detrimental impact human interaction has on the environment. This chapter incorporates in-depth information about technology like conservation drones, air stripping, electric vehicles, treatment ponds, slow sand filters and alternate green energy sources like hydropower.

Chapter 5- The branch of environmental engineering law deals exclusively with complex environmental issues like asbestos baseline survey, soil contamination and assessment and remediation, environmental land use restriction etc. This chapter delineates the local and international environmental laws whose sound knowledge is required to shape environmental engineering technologies. The chapter expounds on the National Environmental Policy Act, Environmental Quality Improvement Act and environmental protocols that have to be followed by environmental engineers.

I would like to share the credit of this book with my editorial team who worked tirelessly on this book. I owe the completion of this book to the never-ending support of my family, who supported me throughout the project.

**Editor**

1

# Introduction to Environmental Engineering

Environmental engineering is a division of engineering that deals with the application of the principles of engineering for the protection of the environment as well as human populations from detrimental human and natural activities. It takes into account the impact that technological advancement has on the quality of the environment. This chapter introduces the reader to these themes and provides an overview of the discipline of environmental engineering.

## Environmental Engineering

Environmental Engineering is the branch of engineering concerned with the application of scientific and engineering principles for protection of human populations from the effects of adverse environmental factors; protection of environments, both local and global, from the potentially deleterious effects of natural and human activities; and improvement of environmental quality.

Environmental engineering can also be described as a branch of applied science and technology that addresses the issues of energy preservation, production asset and control of waste from human and animal activities. Furthermore, it is concerned with finding plausible solutions in the field of public health, such as waterborne diseases, implementing laws which promote adequate sanitation in urban, rural and recreational areas. It involves waste water management, air pollution control, recycling, waste disposal, radiation protection, industrial hygiene, environmental sustainability, public health and environmental engineering law. It also includes studies on the environmental impact of proposed construction projects.

Environmental engineers study the effect of technological advances on the environment. To do so, they conduct studies on hazardous-waste management to evaluate the significance of such hazards, advise on treatment and containment, and develop regulations to prevent mishaps. Environmental engineers design municipal water supply and industrial wastewater treatment systems. They address local and worldwide environmental issues such as the effects of acid rain, global warming, ozone depletion, water pollution and air pollution from automobile exhausts and industrial sources.

At many universities environmental engineering programs are offered at either the department of civil engineering or the department of chemical engineering at engineering faculties. Environmental "civil" engineers focus on hydrology, water resources management, bioremediation, and water treatment plant design. Environmental "chemical" engineers, on the other hand, focus on environmental chemistry, advanced air and water treatment technologies and separation processes.

Additionally, engineers are more frequently obtaining specialized training in law (J.D.) and are utilizing their technical expertise in the practices of environmental engineering law.

Most jurisdictions also impose licensing and registration requirements.

## Development

Ever since people first recognized that their health is related to the quality of their environment, they have applied principles to attempt to improve the quality of their environment. The ancient Indian Harappan civilization utilized early sewers in some cities more than 5000 years ago. The Romans constructed aqueducts to prevent drought and to create a clean, healthful water supply for the metropolis of Rome. In the 15th century, Bavaria created laws restricting the development and degradation of alpine country that constituted the region's water supply.

The field emerged as a separate environmental discipline during the middle third of the 20th century in response to widespread public concern about water and pollution and increasingly extensive environmental quality degradation. However, its roots extend back to early efforts in public health engineering. Modern environmental engineering began in London in the mid-19th century when Joseph Bazalgette designed the first major sewerage system that reduced the incidence of waterborne diseases such as cholera. The introduction of drinking water treatment and sewage treatment in industrialized countries reduced waterborne diseases from leading causes of death to rarities.

In many cases, as societies grew, actions that were intended to achieve benefits for those societies had longer-term impacts which reduced other environmental qualities. One example is the widespread application of the pesticide DDT to control agricultural pests in the years following World War II. While the agricultural benefits were outstanding and crop yields increased dramatically thus reducing world hunger substantially, and malaria was controlled better than it ever had been, numerous species were brought to the verge of extinction due to the impact of the DDT on their reproductive cycles. The story of DDT as vividly told in Rachel Carson's *Silent Spring* (1962) is considered to be the birth of the modern environmental movement and the of the modern field of "environmental engineering."

Conservation movements and laws restricting public actions that would harm the environment have been developed by various societies for millennia. Notable examples are the laws decreeing the construction of sewers in London and Paris in the 19th century and the creation of the U.S. national park system in the early 20th century.

## Scope

## Solid Waste Management

## Environmental Impact Assessment and Mitigation

Scientists have air pollution dispersion models to evaluate the concentration of a pollutant at a receptor or the impact on overall air quality from vehicle exhausts and industrial flue gas stack emissions. To some extent, this field overlaps the desire to decrease carbon dioxide and other greenhouse gas emissions from combustion processes. They apply scientific and engineering principles to evaluate if there are likely to be any adverse impacts to water quality, air quality, habitat quality, flora and fauna, agricultural capacity, traffic impacts, social impacts, ecological impacts, noise impacts, visual (landscape) impacts, etc. If impacts are expected, they then develop mitigation measures to limit or prevent such impacts. An example of a mitigation measure would be the creation of wetlands in a nearby location to mitigate the filling in of wetlands necessary for a road development if it is not possible to reroute the road.

In the United States, the practice of environmental assessment was formally initiated on January 1, 1970, the effective date of the National Environmental Policy Act (NEPA). Since that time, more than 100 developing and developed nations either have planned specific analogous laws or have adopted procedure used elsewhere. NEPA is applicable to all federal agencies in the United States.

## Water Supply and Treatment

Sewage treatment plant, Australia

Engineers and scientists work to secure water supplies for potable and agricultural use. They evaluate the water balance within a watershed and determine the available water supply, the water needed for various needs in that watershed, the seasonal cycles of water movement through the watershed and they develop systems to store, treat, and convey water for various uses. Water is treated to achieve water quality objectives for the end uses. In the case of a potable water supply, water is treated to minimize the risk of infectious disease transmission, the risk of non-infectious illness, and to create a palatable water flavor. Water distribution systems are designed and built to provide adequate water pressure and flow rates to meet various end-user needs such as domestic use, fire suppression, and irrigation.

## Wastewater Treatment

There are numerous wastewater treatment technologies. A wastewater treatment train can consist of a primary clarifier system to remove solid and floating materials, a secondary treatment system consisting of an aeration basin followed by flocculation and sedimentation or an activated sludge system and a secondary clarifier, a tertiary biological nitrogen removal system, and a final disinfection process. The aeration basin/activated sludge system removes organic material by growing bacteria (activated sludge). The secondary clarifier removes the activated sludge from the water. The tertiary system, although not always included due to costs, is becoming more prevalent to remove nitrogen and phosphorus and to disinfect the water before discharge to a surface water stream or ocean outfall.

Water pollution

## Air Pollution Management

Scientists have developed air pollution dispersion models to evaluate the concentration of a pollutant at a receptor or the impact on overall air quality from vehicle exhausts and industrial flue gas stack emissions. To some extent, this field overlaps the desire to decrease carbon dioxide and other greenhouse gas emissions from combustion processes.

## Environmental Protection Agency

The U.S. Environmental Protection Agency (EPA) is one of the many agencies that work with environmental engineers to solve key issues. An important component of

EPA's mission is to protect and improve air, water, and overall environmental quality in order to avoid or mitigate the consequences of harmful effects.

## Ecological Engineering for Sustainable Agriculture in Arid and Semiarid West African Regions

Ecological engineering offers new alternatives for the management of agricultural systems that are more tailored to the ever-changing social and environmental necessities in these regions. This requires managing the complexity of agrosystems, while striving to mimic the functioning of natural ecosystems of West African drylands and taking advantage of traditional practices and local know-how resulting from a long process of adaptation to environmental constraints.

1. Acting on biodiversity. Biodiversity is essential to the productivity of ecosystems and their temporal stability under the impact of external disturbances. Several ecological processes related to biodiversity may be intensified for the benefit of agrosilvopastoral systems: promoting diversity and soil microorganism activity to benefit plants, associating and utilizing the mutual benefits of plants

2. Utilizing organic matter and nutrient cycles. The productivity of agrosystems with low chemical input use in dryland regions is primarily based on efficient organic resource management, and in turn on the nutrient and energy flows they induce. It is thus possible to intervene at several levels: enhancing crop-livestock farming integration to preserve natural resources, restoring the biological activity of soils via specific organic inputs, supplying nutrients to plants locally.

3. Enhancing available water use. Water supplies are limited and irregular in dryland areas. Current management of these supplies—which involves capturing rainwater and surface runoff—could be improved in several ways: adapting to erratic rainfall or drought risks by focusing on: (i) the organization of the farm and community (farm plot patterns in association with the random rainfall distribution, etc.), and on (ii) cropping techniques to reduce crop water needs (plant choices, weeding, etc.), preserving water in crop fields by hampering runoff, accounting for the essential role of trees regarding soil and water in drylands.

4. Managing landscapes and associated ecological processes. Ecological crop pest regulation by their natural enemies is one ecosystem service provided by biodiversity. Better pest management could be considered in association with promoting biodiversity at different scales, e.g. from the plant to the landscape.

## Education

Courses aimed at developing graduates with specific skills in environmental systems or environmental technology are becoming more common and fall into broad classes:

- *Mechanical engineering* courses oriented towards designing machines and mechanical systems for environmental use such as water treatment facilities, pumping stations, garbage segregation plants and other mechanical facilities;
- *Environmental engineering or environmental systems* courses oriented towards a civil engineering approach in which structures and the landscape are constructed to blend with or protect the environment;
- *Environmental chemistry, sustainable chemistry* or *environmental chemical engineering* courses oriented towards understanding the effects (good and bad) of chemicals in the environment. Focus on mining processes, pollutants and commonly also cover biochemical processes;
- *Environmental technology* courses oriented towards producing electronic or electrical graduates capable of developing devices and artifacts able to monitor, measure, model and control environmental impact, including monitoring and managing energy generation from renewable sources.
- Davis, M. L. and D. A. Cornwell, (2006) *Introduction to environmental engineering* (4th ed.) McGraw-Hill ISBN 978-0072424119

## Environmental Engineering Science

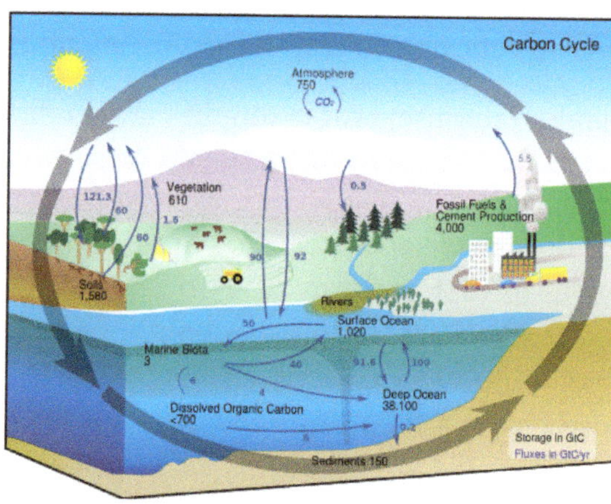


Students in Environmental Engineering Science typically combine scientific studies of the biosphere with mathematical, analytical and design tools found in the engineering fields

Environmental engineering science (EES) is a multidisciplinary field of engineering science that combines the biological, chemical and physical sciences with the field of engineering. This major traditionally requires the student to take basic engineering classes in fields such as thermodynamics, advanced math, computer modeling and simulation and technical classes in subjects such as statics, mechanics, hydrology, and fluid dynamics. As the student progresses, the upper division elective classes define a specific field of study for the student with a choice in a range of science, technology and engineering related classes.

## Difference With Related Fields

As a recently created program, environmental engineering science has not yet been incorporated into the terminology found among environmentally focused professionals. In the few engineering colleges that offer this major, the curriculum shares more classes in common with environmental engineering than it does with environmental science. Typically, EES students follow a similar course curriculum with environmental engineers until their fields diverge during the last year of college. The majority of the environmental engineering students must take classes designed to connect their knowledge of the environment to modern building materials and construction methods. This is meant to direct the environmental engineer into a field where they will more than likely assist in building treatment facilities, preparing environmental impact assessments or helping to mitigate air pollution from specific point sources.

Graduates of Environmental Engineering Science can go on to work on the technical aspects of designing a Living Roof like the one pictured here at the California Academy of the Sciences

Meanwhile, the environmental engineering science student will choose a direction for their career. From the range of electives they have to choose from, these students can move into a fields such as the design of nuclear storage facilities, bacterial bioreactors or environmental policies. These students combine the practical design background of an engineer with the detailed theory found in many of the biological and physical sciences.

## Description at Universities

### Stanford University

The Civil and Environmental Engineering department at Stanford University provides the following description for their program in Environmental Engineering and Science: The Environmental Engineering and Science (EES) program focuses on the chemical and biological processes involved in water quality engineering, water and air pollution, remediation and hazardous substance control, human exposure to pollutants, environmental biotechnology, and environmental protection.

## UC Berkeley

The College of Engineering at UC Berkeley defines Environmental Engineering Science, including the following:

This is a multidisciplinary field requiring an integration of physical, chemical and biological principles with engineering analysis for environmental protection and restoration. The program incorporates courses from many departments on campus to create a discipline that is rigorously based in science and engineering, while addressing a wide variety of environmental issues. Although an environmental engineering option exists within the civil engineering major, the engineering science curriculum provides a more broadly based foundation in the sciences than is possible in civil engineering

## Massachusetts Institute of Technology

At MIT, the major is described in their curriculum, including the following:

The Bachelor of Science in Environmental Engineering Science emphasizes the fundamental physical, chemical, and biological processes necessary for understanding the interactions between man and the environment. Issues considered include the provision of clean and reliable water supplies, flood forecasting and protection, development of renewable and nonrenewable energy sources, causes and implications of climate change, and the impact of human activities on natural cycles

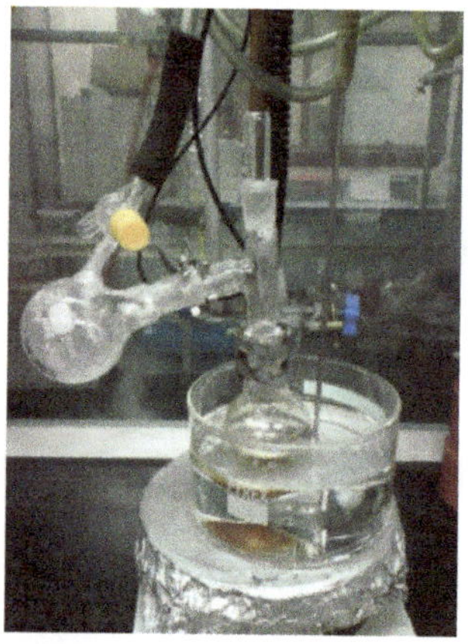

Wet labs are required as part of the lower division curriculum

## Lower Division Coursework

Lower division coursework in this field requires the student to take several laboratory-based classes in calculus-based physics, chemistry, biology, programming and analysis. This is intended to give the student background information in order to introduce them to the engineering fields and to prepare them for more technical information in their upper division coursework.

## Upper Division Coursework

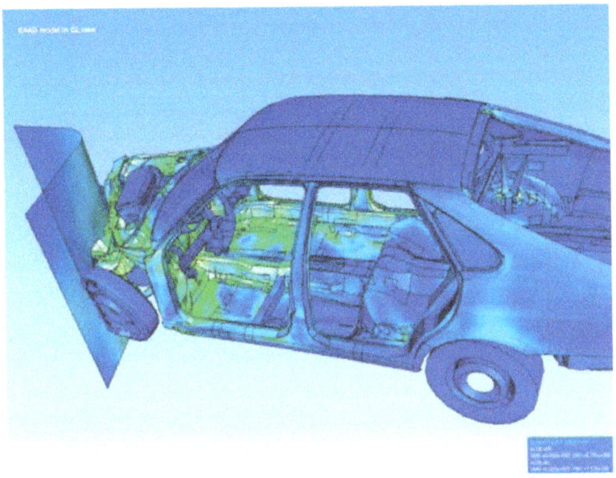

Students learn to integrate their math and statistics with software to perform analysis of physical systems like the Finite Element Analysis shown above

The upper division classes in Environmental Engineering Science prepares the student for work in the fields of engineering and science with coursework in subjects including the following:

- Fluid mechanics
- Mechanics of materials
- Thermodynamics
- Environmental engineering
- Advanced math and statistics
- Geology
- Physical, organic and atmospheric chemistry
- Biochemistry
- Microbiology
- Ecology

## Electives

## Process Engineering

On this track, students are introduced to the fundamental reaction mechanisms in the field of chemical and biochemical engineering.

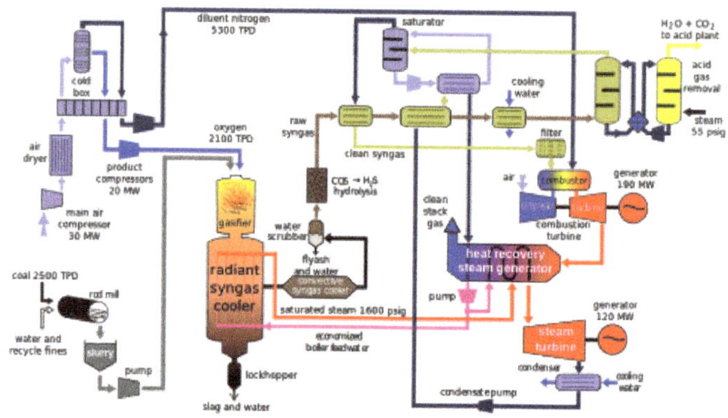


Considering a more environmentally friendly process for coal gasification

## Resource Engineering

For this track, students take classes introducing them to ways to conserve natural resources. This can include classes in water chemistry, sanitation, combustion, air pollution and radioactive waste management.

Using design knowledge to make better wastewater treatment facilities

Designing a safe way to store nuclear waste

## Geoengineering

This examines geoengineering in detail.

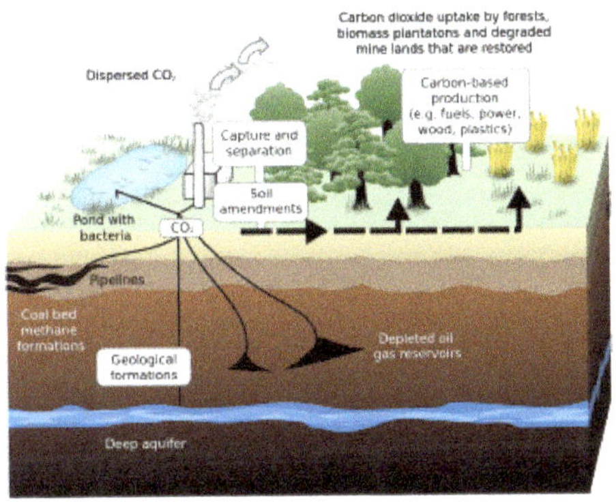


Sequestering carbon from the atmosphere

## Ecology

This prepares the students for using their engineering and scientific knowledge to solve the interactions between plants, animals and the biosphere.

Examining how the harvesting of kelp effects the sea otter population

## Biology

This includes further education about microbial, molecular and cell biology. Classes can include cell biology, virology, microbial and plant biology

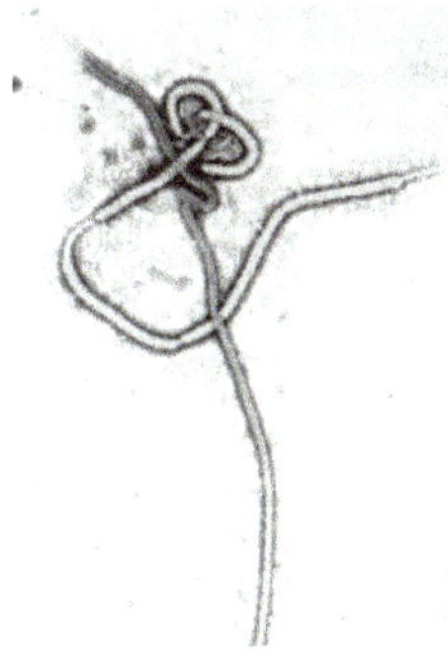

Understanding the way in which viruses function in order to safely sanitize water supplies

## Policy

This covers in more detail ways the environment can be protected through political means. This is done by introducing students to qualitative and quantitative tools in classes such as economics, sociology, political science and energy and resources.

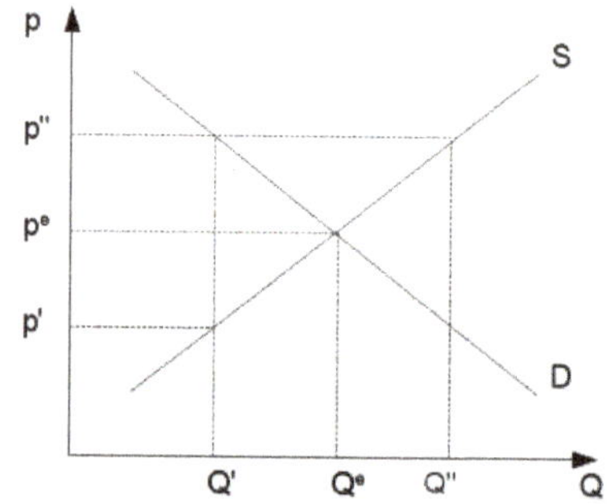


Learning about economics to determine the financial burden it might take to implement an "environmentally friendly" technology

## Post Graduation Work

The multidisciplinary approach in Environmental Engineering Science gives the student expertise in technical fields related to their own personal interest. While some graduates choose to use this major to go to graduate school, students who choose to work often go into the fields of civil and environmental engineering, biotechnology, and research. However, the less technical math, programming and writing background gives the students opportunities to pursue IT work and technical writing.

# Environmental Technology

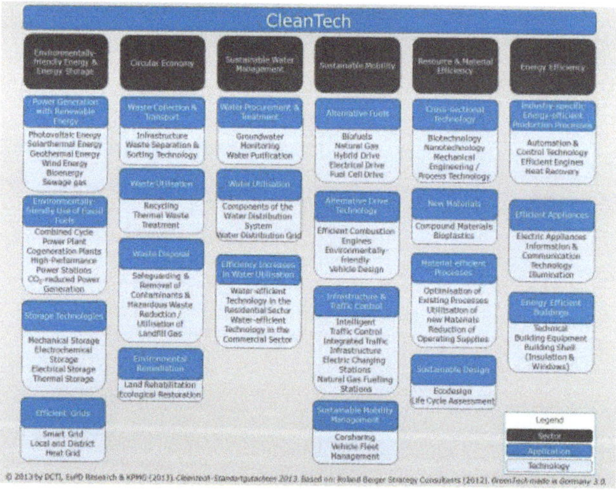


CleanTech

Environmental technology (*envirotech*), green technology (*greentech*) or clean technology (*cleantech*) is the application of one or more of environmental science, green chemistry, environmental monitoring and electronic devices to monitor, model and conserve the natural environment and resources, and to curb the negative impacts of human involvement. The term is also used to describe sustainable energy generation technologies such as photovoltaics, wind turbines, bioreactors, etc. Sustainable development is the core of *environmental technologies*. The term *environmental technologies* is also used to describe a class of electronic devices that can promote sustainable management of resources.

Sustainable urban design and innovation: Photovoltaic ombrière SUDI is an autonomous and mobile station that replenishes energy for electric vehicles using solar energy.

## Examples

- Biofiltration
- Bioreactor
- Bioremediation
- Desalination
- Doubly fed electric machine
- Energy conservation
- Energy saving modules
- Electric vehicles
- Wave energy
- Green computing
- Hydroelectricity
- Wind power
- Wind turbine

- Hydrogen fuel cell
- Ocean thermal energy conversion
- Solar power
- Photovoltaic
- Thermal depolymerization
- Composting toilet
- Pyrolysis

## Renewable Energy

Renewable energy is the energy that can be replenished easily. For years we have been using sources like wood, sun, water, etc. for means for producing energy. Energy that can be produced by natural objects like wood, sun, wind, etc. is considered to be renewable.

A view across a reverse osmosis desalination plant in Nepal.

## Water Purification

Water purification: The whole idea/concept of having dirt/germ/pollution free water flowing throughout the environment. Many other phenomena lead from this concept of purification of water. Water pollution is the main enemy of this concept, and various campaigns and activists have been organized around the world to help purify water.

## Air Purification

Air purification: Basic and common green plants can be grown indoors to keep air fresh because all plants remove $CO_2$ and convert it into oxygen. The best examples are:

*Dypsis lutescens*, *Sansevieria trifasciata*, and *Epipremnum aureum*.

## Sewage Treatment

Sewage treatment is conceptually similar to water purification. Sewage treatments are very important as they purify water per levels of its pollution. The most polluted water is not used for anything, and the least polluted water is supplied to places where water is used affluently. It may lead to various other concepts of environmental protection, sustainability etc.

## Environmental Remediation

Environmental remediation is the removal of pollutants or contaminants for the general protection of the environment. This is accomplished by various chemical, biological, and bulk movements. (encyclopedia of medical concepts)

## Solid Waste Management

Net Zero Court zero emissions office building prototype in St. Louis, Missouri

Solid waste management is the purification, consumption, reuse, disposal and treatment of solid waste that is undertaken by the government or the ruling bodies of a city/town.

## eGain Forecasting

Egain forecasting is a method using forecasting technology to predict the future weather's impact on a building. By adjusting the heat based on the weather forecast, the system eliminates redundant use of heat, thus reducing the energy consumption and the emission of greenhouse gases.

## Energy Conservation

Energy conservation is the utilization of devices that require smaller amounts of energy in order to reduce the consumption of electricity. Reducing the use of electricity causes less fossil fuels to be burned to provide that electricity.

## Alternative and Clean Power

The Tesla Roadster was the first all-electric sports car for sale and in serial production. It can be plugged into conventional outlets and can be charged fully or partially on renewable energy, including solar, hydroelectric, geothermal or wind power.

## Principles:

- Green syndicalism
- Sustainability
- Sustainable design
- Sustainable engineering

Scientists continue to search for clean energy alternatives to our current power production methods. Some technologies such as anaerobic digestion produce renewable energy from waste materials. The global reduction of greenhouse gases is dependent on the adoption of energy conservation technologies at industrial level as well as this clean energy generation. That includes using unleaded gasoline, solar energy and alternative fuel vehicles, including plug-in hybrid and hybrid electric vehicles.

Since industrial use of energy accounts for 51% of worldwide energy usage improving energy efficiency in this field is a top priority for environmental technology companies around the globe. Advanced energy efficient electric motor (and electric generator) technology that are cost effective to encourage their application, such as the brushless wound-rotor doubly fed electric machine and energy saving module, can reduce the amount of carbon dioxide ($CO_2$) and sulfur dioxide ($SO_2$) that would otherwise be introduced to the atmosphere, if electricity is generated using fossil fuels. Greasestock is an event held yearly in Yorktown Heights, New York which is one of the largest showcases of environmental technology in the United States.

## Reversed Chemical Energy to Mechanical Energy

Reversed Chemistry: In Philosophical Era Still, waiting for Practical, What ever we use as fuel to run engines, is cool to hot, life on earth is result of coolness....... such coolness is also available in fossils, as so in fossil fuel, when we spark fire on it is flame, and there is also reversed chemistry and formula, as putting cool spark on hot chemical combinations. like in engine put carbide and water reaction or on carbide like or more hot chemical put little water drop or more cool chemical like fossil fuel it will make pressure for engines..there are more chemical formula like this, for example reaction on water in nitric acid

there are things made on heat process like iron, steel, zinc etc., it also may be reversed making things on freezing or cooling process, like cement and water reaction make strong chemical molecules, or glue or super glue on heat they do not melt, but iron, steel and zinc melt on heat, so super glue like chemical formula with more chemical work will be so strong if it will be processed from high cool temperature (minus temperature)

## Education

Courses aimed at developing graduates with specific skills in environmental systems or environmental technology are becoming more common and fall into three broads classes:

- *Environmental Engineering or Environmental Systems* courses oriented towards a civil engineering approach in which structures and the landscape are constructed to blend with or protect the environment;
- *Environmental chemistry*, *sustainable chemistry* or *environmental chemical engineering* courses oriented towards understanding the effects (good and bad) of chemicals in the environment. Such awards can focus on mining processes, pollutants and commonly also cover biochemical processes;
- *Environmental technology* courses oriented towards producing electronic, electrical or electrotechnology graduates capable of developing devices and artefacts able to monitor, measure, model and control environmental impact, including monitoring and managing energy generation from renewable sources, and developing novel energy generation technologies.

## References


- Turner, D.B. (1994). Workbook of atmospheric dispersion estimates: an introduction to dispersion modeling (2nd ed.). CRC Press. ISBN 1-56670-023-X.
- Davis, M. L. and D. A. Cornwell, (2006) Introduction to environmental engineering (4th ed.) McGraw-Hill ISBN 978-0072424119
- Tchobanoglous, G.; Burton, F.L. & Stensel, H.D. (2003). Wastewater Engineering (Treatment Disposal Reuse) / Metcalf & Eddy, Inc. (4th ed.). McGraw-Hill Book Company. ISBN 0-07-041878-0.

- Beychok, M.R. (2005). Fundamentals Of Stack Gas Dispersion (4th ed.). author-published. ISBN 0-9644588-0-2.
- D. Masse; JL. Chotte; E. Scopel (2015). "Ecological engineering for sustainable agriculture in arid and semiarid West African regions". Fiche thématique du CSFD (11): 2
- "Architecture and Engineering Occupations : Occupational Outlook Handbook : U.S. Bureau of Labor Statistics". Bls.gov. 2012-03-29. Retrieved 2013-07-01.
- "Funding - Environmental Engineering - US National Science Foundation (NSF)". nsf.gov. Retrieved 2013-07-01.

# 2

# Major Areas of Study of Environmental Engineering

This chapter expounds on the major areas of environmental engineering including vehicle emissions control, ecological engineering, sanitation, industrial waste treatment and public health. These areas are important as they form the crux of the issues that are tackled by environmental engineering. The topics discussed in the chapter are of great importance to broaden the existing knowledge on environmental engineering.

## Environmental Impact Assessment

*Environmental assessment*' (EA) is the term used for the assessment of the environmental consequences (positive and negative) of a plan, policy, program, or concrete projects prior to the decision to move forward with the proposed action. In this context, the term 'environmental impact assessment' (EIA) is usually used when applied to concrete projects by individuals or companies and the term 'strategic environmental assessment' (SEA) applies to policies, plans and programmes most often proposed by organs of state (Fischer, 2016). Environmental assessments may be governed by rules of administrative procedure regarding public participation and documentation of decision making, and may be subject to judicial review.

The purpose of the assessment is to ensure that decision makers consider the environmental impacts when deciding whether or not to proceed with a project. The International Association for Impact Assessment (IAIA) defines an environmental impact assessment as "the process of identifying, predicting, evaluating and mitigating the biophysical, social, and other relevant effects of development proposals prior to major decisions being taken and commitments made." EIAs are unique in that they do not require adherence to a predetermined environmental outcome, but rather they require decision makers to account for environmental values in their decisions and to justify those decisions in light of detailed environmental studies and public comments on the potential environmental impacts.

CASI Global, New York (http://casiglobal.us/) The Global Certification body for CSR & Sustainability goes one step further in this definition to include financial and supply chain effects. Any effects of the projects supply chain on the environment, ecology, human resources or country resources are to be included in the impact analysis. Most

Global Fellows of CASI NY have to write a paper on the same and are authorized to undertake such projects in their country.

Engineering and consulting companies work hand in hand as contractors for mining, energy, oil & gas companies executing EIAs. Companies operating globally such as Arcadis, Royal HaskoningDHV, Golder Associates, Amec Foster Wheeler, Schlumberger Water Services (a Schlumberger company), ERM are an example of a much bigger pool of expertise globally. These contractors are the ones not only in charge of preparing an EIA study but most importantly getting these studies approved by each country government offices prior to the execution of a project. Each country will also have its own local contractors offering the same kind of service hence breaking out monopolies by increasing the supply of EIAs execution consultants.

## History

Environmental impact assessments commenced in the 1960s, as part of increasing environmental awareness. EIAs involved a technical evaluation intended to contribute to more objective decision making. In the United States, environmental impact assessments obtained formal status in 1969, with enactment of the National Environmental Policy Act. EIAs have been used increasingly around the world. The number of "Environmental Assessments" filed every year "has vastly overtaken the number of more rigorous Environmental Impact Statements (EIS)." An Environmental Assessment is a "mini-EIS designed to provide sufficient information to allow the agency to decide whether the preparation of a full-blown Environmental Impact Statement (EIS) is necessary." EIA is an activity that is done to find out the impact that would be done before development will occur.

## Methods

General and industry specific assessment methods are available including:

- *Industrial products* - Product environmental life cycle analysis (LCA) is used for identifying and measuring the impact of industrial products on the environment. These EIAs consider activities related to extraction of raw materials, ancillary materials, equipment; production, use, disposal and ancillary equipment.
- *Genetically modified plants* - Specific methods available to perform EIAs of genetically modified organisms include GMP-RAM and INOVA.
- *Fuzzy logic* - EIA methods need measurement data to estimate values of impact indicators. However, many of the environment impacts cannot be quantified, e.g. landscape quality, lifestyle quality and social acceptance. Instead information from similar EIAs, expert judgment and community sentiment are

employed. Approximate reasoning methods known as fuzzy logic can be used. A fuzzy arithmetic approach has also been proposed and implemented using a software tool (TDEIA). More information can be found at ARAI web site.

## Follow-up

At the end of the project, an audit evaluates the accuracy of the EIA by comparing actual to predicted impacts. The objective is to make future EIAs more valid and effective. Two primary considerations are:

- *Scientific* - to examine the accuracy of predictions and explain errors
- *Management* - to assess the success of mitigation in reducing impacts

Audits can be performed either as a rigorous assessment of the null hypothesis or with a simpler approach comparing what actually occurred against the predictions in the EIA document.

After an EIA, the precautionary and polluter pays principles may be applied to decide whether to reject, modify or require strict liability or insurance coverage to a project, based on predicted harms.

The Hydropower Sustainability Assessment Protocol is a sector specific method for checking the quality of Environmental and Social assessments and management plans.

## Around The World

### Australia

The history of EIA in Australia could be linked to the enactment of the U.S. National Environment Policy Act (NEPA) in 1970, which made the preparation of environmental impact statements a requirement. In Australia, one might say that the EIA procedures were introduced at a State Level prior to that of the Commonwealth (Federal), with a majority of the states having divergent views to the Commonwealth. One of the pioneering states was New South Wales, whose State Pollution Control Commission issued EIA guidelines in 1974. At a Commonwealth (Federal) level, this was followed by passing of the Environment Protection (Impact of Proposals) Act in 1974. The Environment Protection and Biodiversity Conservation Act 1999 (EPBC) superseded the Environment Protection (Impact of Proposals) Act 1974 and is the current central piece for EIA in Australia on a Commonwealth (Federal) level. An important point to note is that this Commonwealth Act does not affect the validity of the States and Territories environmental and development assessments and approvals; rather the EPBC runs as a parallel to the State/Territory Systems. Overlap between federal and state requirements is addressed via bilateral agreements or one off accreditation of state processes, as provided for in the EPBC Act.

## The Commonwealth Level

The EPBC Act provides a legal framework to protect and manage nationally and internationally important flora, fauna, ecological communities and heritage places-defined in the EPBC Act as matters of 'national environmental significance'. Following are the eight matters of 'national environmental significance' to which the EPBC ACT applies:

- World Heritage sites
- National Heritage places
- RAMSAR wetlands of international significance
- Listed threatened species and ecological communities
- Migratory species protected under international agreements
- The Commonwealth marine environment
- Nuclear actions (including uranium mining)
- National Heritage.

In addition to this, the EPBC Act aims at providing a streamlined national assessment and approval process for activities. These activities could be by the Commonwealth, or its agents, anywhere in the world or activities on Commonwealth land; and activities that are listed as having a 'significant impact' on matters of 'national environment significance'.

The EPBC Act comes into play when a person (a 'proponent') wants an action (often called a 'proposal' or 'project') assessed for environmental impacts under the EPBC Act, he or she must refer the project to the Department of Environment, Water, Heritage and the Arts (Australia). This 'referral' is then released to the public, as well as relevant state, territory and Commonwealth ministers, for comment on whether the project is likely to have a significant impact on matters of national environmental significance. The Department of Environment, Water, Heritage and the Arts assess the process and makes recommendation to the minister or the delegate for the feasibility. The final discretion on the decision remains of the minister, which is not solely based on matters of 'national environmental significance' but also the consideration of social and economic impact of the project.

The Australian Government environment minister cannot intervene in a proposal if it has no significant impact on one of the eight matters of 'national environmental significance' despite the fact that there may be other undesirable environmental impacts. This is primarily due to the division of powers between the States and the Federal government and due to which the Australian Government environment minister cannot overturn a state decision.

There are strict civil and criminal penalties for the breach of EPBC Act. Depending on the kind of breach, civil penalty (maximum) may go up to $550,000 for an individual and $5.5 million for a body corporate, or for criminal penalty (maximum) of seven years imprisonment and/or penalty of $46,200.

## The State and Territory Level

Australian Capital Territory (ACT)

EIA provisions within Ministerial Authorities in the ACT are found in the Chapters 7 and 8 of the *Planning and Development Act 2007* (ACT). EIA in ACT was previously administered with the help of Part 4 of the Land (Planning and Environment) Act 1991 (Land Act) and Territory Plan (plan for land-use). Note that some EIA may occur in the ACT on Commonwealth land under the EPBC Act (Cth). Further provisions of the *Australian Capital Territory (Planning and Land Management) Act 1988* (Cth) may also be applicable particularly to national land and "designated areas".

New South Wales (NSW)

In New South Wales, the Environment Planning Assessment Act 1979 (EPA) establishes three pathways for EIA. The first is under Part 5.1 of the EPAA, which provides for EIA of 'State Significant Infrastructure' projects. (From June 2011, this Part replaced Part 3A, which previously covered EIA of major projects). The second is under Part 4 of the Act dealing with development control. If a project does not require approval under Part 3A or Part 4 it is then potentially captured by the third pathway, Part 5 dealing with environment impact assessment.

Northern Territory (NT)

The EIA process in Northern Territory is chiefly administered under the Environmental Assessment Act (EAA). Although EAA is the primary tool for EIA in Northern Territory, there are further provisions for proposals in the Inquiries Act 1985 (NT).

Queensland (QLD)

There are four main EIA processes in Queensland. Firstly, under the Integrated Planning Act 1997 (IPA) for development projects other than mining. Secondly, under the Environmental Protection Act 1994 (EP Act) for some mining and petroleum activities. Thirdly, under the State Development and Public Works Organization Act 1971 (State Development Act) for 'significant projects'. Finally, Environment Protection and Biodiversity Conservation Act 1999 (Cth) for 'controlled actions'.

South Australia (SA)

The local governing tool for EIA in South Australia is the Development Act 1993. There are three levels of assessment possible under the Act in the form of an environment impact statement (EIS), a public environmental report (PER) or a Development Report (DR).

Tasmania (TAS)

In Tasmania, an integrated system of legislation is used to govern development and approval process, this system is a mixture of the Environmental Management and Pollution Control Act 1994 (EMPCA), Land Use Planning and Approvals Act 1993 (LUPAA), State Policies and Projects Act 1993 (SPPA), and Resource Management and Planning Appeals Tribunal Act 1993.

Victoria (VIC)

The EIA process in Victoria is intertwined with the Environment Effects Act 1978 and the Ministerial Guidelines for Assessment of Environmental Effects (made under the s. 10 of the EE Act).

Western Australia (WA)

The Environmental Protection Act 1986 (Part 4) provides the legislative framework for the EIA process in Western Australia. The EPA Act oversees the planning and development proposals and assesses their likely impacts on the environment.

## Canada

In *Friends of the Oldman River Society v. Canada (Minister of Transportation)*,(SCC 1992) La Forest J of the Supreme Court of Canada described environmental impact assessment in terms of the proper scope of federal jurisdiction with respect to environments matters,

"Environmental impact assessment is, in its simplest form, a planning tool that is now generally regarded as an integral component of sound decision-making."

Supreme Court Justice La Forest cited (Cotton, Emond & 1981 245), "The basic concepts behind environmental assessment are simply stated: (1) early identification and evaluation of all potential environmental consequences of a proposed undertaking; (2) decision making that both guarantees the adequacy of this process and reconciles, to the greatest extent possible, the proponent's development desires with environmental protection and preservation."

La Forest referred to (Jeffrey 1989, 1.2,1.4) and (Emond 1978, p. 5) who described "... environmental assessments as a planning tool with both an information-gathering and a decision-making component" that provide "...an objective basis for granting or denying approval for a proposed development."

Justice La Forest addressed his concerns about the implications of Bill C-45 regarding public navigation rights on lakes and rivers that would contradict previous cases.(La Forest & 1973 178-80)

The Canadian Environmental Assessment Act 2012 (CEAA 2012) "and its regulations establish the legislative basis for the federal practice of environmental assessment in most regions of Canada." CEAA 2012 came into force July 6, 2012 and replaces the former *Canadian Environmental Assessment Act* (1995). EA is defined as a planning tool to identify, understand, assess and mitigate, where possible, the environmental effects of a project.

"The purposes of this Act are: (a) to protect the components of the environment that are within the legislative authority of Parliament from significant adverse environmental effects caused by a designated project; (b) to ensure that designated projects that require the exercise of a power or performance of a duty or function by a federal authority under any Act of Parliament other than this Act to be carried out, are considered in a careful and precautionary manner to avoid significant adverse environmental effects; (c) to promote cooperation and coordinated action between federal and provincial governments with respect to environmental assessments; (d) to promote communication and cooperation with aboriginal peoples with respect to environmental assessments; (e) to ensure that opportunities are provided for meaningful public participation during an environmental assessment; (f) to ensure that an environmental assessment is completed in a timely manner; (g) to ensure that projects, as defined in section 66, that are to be carried out on federal lands, or those that are outside Canada and that are to be carried out or financially supported by a federal authority, are considered in a careful and precautionary manner to avoid significant adverse environmental effects; (h) to encourage federal authorities to take actions that promote sustainable development in order to achieve or maintain a healthy environment and a healthy economy; and (i) to encourage the study of the cumulative effects of physical activities in a region and the consideration of those study results in environmental assessments."

*Canadian Environmental Assessment Act*

## Opposition

Environmental Lawyer Dianne Saxe argued that the CEAA 2012 "allows the federal government to create mandatory timelines for assessments of even the largest and most important projects, regardless of public opposition." (Saxe 2012)

"Now that federal environmental assessments are gone, the federal government will only assess very large, very important projects. But it's going to do them in a hurry."

*Dianne Saxe*

On 3 August 2012 the Canadian Environmental Assessment Agency nine "designated projects" with their timelines: Enbridge Northern Gateway Pipeline Joint Review Panel (JRP) 18 months; Marathon Platinum Group Metals and Copper Mine Project (JRP): 13 months; Site C Clean Energy Project (JRP) 8.5 months; Deep Geologic Repository Project (JRP) 17 months; Enbridge Northern Gateway Project (JRP) 18 months; Jackpine Mine Expansion Project (JRP) 11.5 months; Pierre River Mine Project: 8 months;

New Prosperity Gold-Copper Mine Project (JRP) 7.5 months; Frontier Oil Sands Mine Project (JRP)8.5 months; EnCana/Cenovus Shallow Gas Infill Project (JRP) 5 months.

Saxe compares these timelines with environmental assessments for the Mackenzie Valley Pipeline. Thomas R. Berger, Royal Commissioner of the Mackenzie Valley Pipeline Inquiry (9 May 1977), worked extremely hard to ensure that industrial development on Aboriginal people's land resulted in benefits to those indigenous people.

On 22 April 2013, Official Opposition Environment critic Megan Leslie issued a statement claiming that the federal government's recent changes to "fish habitat protection, the Navigable Waters Protection Act and the Canadian Environmental Assessment Act", along with gutting existing laws and making cuts to science and research, "will be disastrous, not only for the environment, but also for Canadians' health and economic prosperity." On 26 September 2012, Leslie argued that with the changes to the Canadian Environmental Assessment Act that came into effect 6 July 2012, "seismic testing, dams, wind farms and power plants" no longer required any federal environmental assessment. She also claimed that because the CEAA 2012—which she claimed was rushed through Parliament—dismantled the CEAA 1995, the Oshawa ethanol plant project would no longer have a full federal environmental assessment. Mr. Peter Kent (Minister of the Environment) explained that the CEAA 2012 "provides for the Government of Canada and the Environmental Assessment Agency to focus on the large and most significant projects that are being proposed across the country." The 2,000 to 3,000-plus smaller screenings that were in effect under CEAA 1995 became the "responsibility of lower levels of government but are still subject to the same strict federal environmental laws." Anne Minh-Thu Quach, MP for Beauharnois—Salaberry, QC, argued that the mammoth budget bill dismantled 50 years of environmental protection without consulting Canadians about the "colossal changes they are making to environmental assessments." She claimed that the federal government is entering into "limited consultations, by invitation only, months after the damage was done."

## China

The Environmental Impact Assessment Law (EIA Law) requires that an environmental impact assessment be completed prior to project construction. However, if a developer completely ignores this requirement and builds a project without submitting an environmental impact statement, the only penalty is that the environmental protection bureau (EPB) may require the developer to do a make-up environmental assessment. If the developer does not complete this make-up assessment within the designated time, only then is the EPB authorized to fine the developer. Even so, the possible fine is capped at a maximum of about US$25,000, a fraction of the overall cost of most major projects. The lack of more stringent enforcement mechanisms has resulted in a significant percentage of projects not completing legally required environmental impact assessments prior to construction.

China's State Environmental Protection Administration (SEPA) used the legislation to halt 30 projects in 2004, including three hydro-power plants under the Three Gorges Project Company. Although one month later (Note as a point of reference, that the typical EIA for a major project in the USA takes one to two years.), most of the 30 halted projects resumed their construction, reportedly having passed the environmental assessment, the fact that these key projects' construction was ever suspended was notable.

A joint investigation by SEPA and the Ministry of Land and Resources in 2004 showed that 30-40% of the mining construction projects went through the procedure of environment impact assessment as required, while in some areas only 6-7% did so. This partly explains why China has witnessed so many mining accidents in recent years.

SEPA alone cannot guarantee the full enforcement of environmental laws and regulations, observed Professor Wang Canfa, director of the centre to help environmental victims at China University of Political Science and Law. In fact, according to Wang, the rate of China's environmental laws and regulations that are actually enforced is estimated at barely 10%.

## Egypt

Environmental Impact Assessment (EIA) EIA is implemented in Egypt under the umbrella of the Ministry of state for environmental affairs. The Egyptian Environmental Affairs Agency (EEAA) is responsible for the EIA services.

In June 1997, the responsibility of Egypt's first full-time Minister of State for Environmental Affairs was assigned as stated in the Presidential Decree no.275/1997. From thereon, the new ministry has focused, in close collaboration with the national and international development partners, on defining environmental policies, setting priorities and implementing initiatives within a context of sustainable development.

According to the Law 4/1994 for the Protection of the Environment, the Egyptian Environmental Affairs Agency (EEAA) was restructured with the new mandate to substitute the institution initially established in 1982. At the central level, EEAA represents the executive arm of the Ministry.

The purpose of EIA is to ensure the protection and conservation of the environment and natural resources including human health aspects against uncontrolled development. The long-term objective is to ensure a sustainable economic development that meets present needs without compromising future generations ability to meet their own needs. EIA is an important tool in the integrated environmental management approach.

EIA must be performed for new establishments or projects and for expansions or renovations of existing establishments according to the Law for the Environment.

## EU

There is a wide range of instruments in the Environmental policy of the European Union. Among them the European Union has established a mix of mandatory and discretionary procedures to assess environmental impacts. European Union Directive (85/337/EEC) on Environmental Impact Assessments (known as the *EIA Directive*) was first introduced in 1985 and was amended in 1997. The directive was amended again in 2003, following EU signature of the 1998 Aarhus Convention, and once more in 2009. The initial Directive of 1985 and its three amendments have been codified in Directive 2011/92/EU of 13 December 2011. In 2001, the issue was enlarged to the assessment of plans and programmes by the so-called *Strategic Environmental Assessment (SEA) Directive* (2001/42/EC), which is now in force. Under the EU directive, an EIA must provide certain information to comply. There are seven key areas that are required:

1. Description of the project
    - Description of actual project and site description
    - Break the project down into its key components, i.e. construction, operations, decommissioning
    - For each component list all of the sources of environmental disturbance
    - For each component all the inputs and outputs must be listed, e.g., air pollution, noise, hydrology
2. Alternatives that have been considered
    - Examine alternatives that have been considered
    - Example: in a biomass power station, will the fuel be sourced locally or nationally?
3. Description of the environment
    - List of all aspects of the environment that may be affected by the development
    - Example: populations, fauna, flora, air, soil, water, humans, landscape, cultural heritage
    - This section is best carried out with the help of local experts, e.g. the RSPB in the UK
4. Description of the significant effects on the environment
    - The word significant is crucial here as the definition can vary

- 'Significant' must be defined
- The most frequent method used here is use of the Leopold matrix
- The matrix is a tool used in the systematic examination of potential interactions
- Example: in a windfarm development a significant impact may be collisions with birds

5. Mitigation
    - This is where EIA is most useful
    - Once section 4 is complete, it is obvious where impacts are greatest
    - Using this information ways to avoid negative impacts should be developed
    - Best working with the developer with this section as they know the project best
    - Using the windfarm cxamplc again construction could be out of bird nesting seasons
6. Non-technical summary (EIS)
    - The EIA is in the public domain and be used in the decision making process
    - It is important that the information is available to the public
    - This section is a summary that does not include jargon or complicated diagrams
    - It should be understood by the informed lay-person
7. Lack of know-how/technical difficulties
    - This section is to advise any areas of weakness in knowledge
    - It can be used to focus areas of future research
    - Some developers see the EIA as a starting block for poor environmental management

## Annexed Projects

All projects are either classified as Annex 1 or Annex 2 projects. Those lying in Annex 1 are large scale developments such as motorways, chemical works, bridges, powerstations etc. These always require an EIA under the Environmental Impact Assessment Directive (85,337,EEC as amended). Annex 2 projects are smaller in scale than those referred to in Annex 1. Member States must determine whether these project shall be made subject to an assessment subject to a set of criteria set out in Annex 3 of codified Directive 2011/92/EU.

## The Netherlands

EIA was implemented in Dutch legislation on September 1, 1987. The categories of projects that require an EIA are summarised in Dutch legislation, the Wet milieubeheer. The use of thresholds for activities makes sure that EIA is obligatory for those activities that may have considerable impacts on the environment.

For projects and plans that fit these criteria, an EIA report is required. The EIA report defines a.o. the proposed initiative, it makes clear the impact of that initiative on the environment and compares this with the impact of possible alternatives with less a negative impact.

## Hong Kong

EIA in Hong Kong, since 1998, is regulated by the *Environmental Impact Assessment Ordinance 1997*.

The original proposal to construct the Lok Ma Chau Spur Line overground across the Long Valley failed to get through EIA, and the Kowloon–Canton Railway Corporation had to change its plan and build the railway underground. In April 2011, the EIA of the Hong Kong section of the Hong Kong-Zhuhai-Macau Bridge was found to have breached the ordinance, and was declared unlawful. The appeal by the government was allowed in September 2011. However, it was estimated that this EIA court case had increased the construction cost of the Hong Kong section of the bridge by HK$6.5 billion in money-of-the-day prices.

## India

The Ministry of Environment, Forests and Climate Change (MoEFCC) of India has been in a great effort in Environmental Impact Assessment in India. The main laws in action are the Water Act(1974), the Indian Wildlife (Protection) Act (1972), the Air (Prevention and Control of Pollution) Act (1981) and the Environment (Protection) Act (1986),Biological Diversity Act(2002). The responsible body for this is the Central Pollution Control Board. Environmental Impact Assessment (EIA) studies need a significant amount of primary and secondary environmental data. Primary data are those collected in the field to define the status of the environment (like air quality data, water quality data etc.). Secondary data are those collected over the years that can be used to understand the existing environmental scenario of the study area. The environmental impact assessment (EIA) studies are conducted over a short period of time and therefore the understanding of the environmental trends, based on a few months of primary data, has limitations. Ideally, the primary data must be considered along with the secondary data for complete understanding of the existing environmental status of the area. In many EIA studies, the secondary data needs could be as high as 80% of the total data requirement. EIC is the repository of one stop secondary data source for environmental impact assessment in India.

The Environmental Impact Assessment (EIA) experience in India indicates that the lack of timely availability of reliable and authentic environmental data has been a major bottle neck in achieving the full benefits of EIA. The environment being a multi-disciplinary subject, a multitude of agencies are involved in collection of environmental data. However, no single organization in India tracks available data from these agencies and makes it available in one place in a form required by environmental impact assessment practitioners. Further, environmental data is not available in enhanced forms that improve the quality of the EIA. This makes it harder and more time-consuming to generate environmental impact assessments and receive timely environmental clearances from regulators. With this background, the Environmental Information Centre (EIC) has been set up to serve as a professionally managed clearing house of environmental information that can be used by MoEF, project proponents, consultants, NGOs and other stakeholders involved in the process of environmental impact assessment in India. EIC caters to the need of creating and disseminating of organized environmental data for various developmental initiatives all over the country.

EIC stores data in GIS format and makes it available to all environmental impact assessment studies and to EIA stakeholders in a cost effective and timely manner. So that we can manage that in different proportions such as remedy measures etc.,

## Korea, South

Recycling culture and policy Ministry of Environment

## Malaysia

In Malaysia, Section 34A, Environmental Quality Act, 1974 requires developments that have significant impact to the environment are required to conduct the Environmental impact assessment.

## Nepal

In Nepal, EIA has been integrated in major development projects since the early 1980s. In the planning history of Nepal, the sixth plan (1980–85), for the first time, recognized the need for EIA with the establishment of Environmental Impact Study Project (EISP) under the Department of Soil Conservation in 1982 to develop necessary instruments for integration of EIA in infrastructure development projects. However, the government of Nepal enunciated environment conservation related policies in the seventh plan (NPC, 1985–1990). To enforce this policy and make necessary arrangements, a series of guidelines were developed, thereby incorporating the elements of environmental factors right from the project formulation stage of the development plans and projects and to avoid or minimize adverse effects on the ecological system. In addition, it has also emphasized that EIAs of industry, tourism, water resources, transportation, urbanization, agriculture, forest and other developmental projects be conducted.

In Nepal, the government's Environmental Impact Assessment Guideline of 1993 inspired the enactment of the Environment Protection Act (EPA) of 1997 and the Environment Protection Rules (EPR) of 1997 (EPA and EPR have been enforced since 24 and 26 June 1997 respectively in Nepal) to internalizing the environmental assessment system. The process institutionalized the EIA process in development proposals and enactment, which makes the integration of IEE and EIA legally binding to the prescribed projects. The projects, requiring EIA or IEE, are included in Schedules 1 and 2 of the EPR, 1997 (GoN/MoLJPA 1997). Progresses were made in the Environmental protection issue during the 8th five-year plan (1992–1997). The following development in Environmental protection were achieved during that time:

- Formulation of Environmental Protection Act 1997
- Establishment of Ministry of Environment
- Development of National Environmental Policies and Action Plan, EIA guidelines developed
- Consideration of environmental concerns in hydropower projects
- Development of industrial, irrigation and agricultural policies that undertook environmental concerns

## New Zealand

In New Zealand, EIA is usually referred to as *Assessment of Environmental Effects* (AEE). The first use of EIA's dates back to a Cabinet minute passed in 1974 called Environmental Protection and Enhancement Procedures. This had no legal force and only related to the activities of government departments. When the Resource Management Act was passed in 1991, an EIA was required as part of a resource consent application. Section 88 of the Act specifies that the AEE must include "such detail as corresponds with the scale and significance of the effects that the activity may have on the environment". While there is no duty to consult any person when making a resource consent application (Sections 36A and Schedule 4), proof of consultation is almost certain required by local councils when they decide whether or not to publicly notify the consent application under Section 93.

## Russian Federation

As of 2004, the state authority responsible for conducting the State EIA in Russia has been split between two Federal bodies: 1) Federal service for monitoring the use of natural resources – a part of the Russian Ministry for Natural Resources and Environment and 2) Federal Service for Ecological, Technological and Nuclear Control. The two main pieces of environmental legislation in Russia are: The Federal Law 'On Ecological Expertise, 1995 and the 'Regulations on Assessment of Impact from Intended Business and Other Activity on Environment in the Russian Federation, 2000.

## Federal Service for Monitoring the Use of Natural Resources

In 2006, the parliament committee on ecology in conjunction with the Ministry for Natural Resources and Environment, created a working group to prepare a number of amendments to existing legislation to cover such topics as stringent project documentation for building of potentially environmentally damaging objects as well as building of projects on the territory of protected areas. There has been some success in this area, as evidenced from abandonment of plans to construct a gas pipe-line through the only remaining habitat of the critically endangered Amur leopard in the Russian Far East.

## Federal Service for Ecological, Technological and Nuclear Control

The government's decision to hand over control over several important procedures, including state EIA in the field of all types of energy projects, to the Federal Service for Ecological, Technological and Nuclear Control had caused a major controversy and criticism from environmental groups that blamed the government for giving nuclear power industry control over the state EIA.

Not surprisingly the main problem concerning State EIA in Russia is the clear differentiation of jurisdiction between the two above-mentioned Federal bodies.

## Sri Lanka

Environmental Impact Assessments

One popular approach to assist in smart growth in democratic countries is for law-makers to require prospective developers to prepare environmental impact assessments of their plans as a condition for state and/or local governments to go for Environmental Impact Assessments.

These reports often indicate how significant impacts the development generates can be mitigated, usually at developer expense. These assessments are frequently controversial. Conservationists, neighborhood advocacy groups and NIMBYs are often skeptical about such impact reports, even when prepared by independent agencies and approved by decision makers rather than promoters. Conversely, developers sometimes strongly resist requirements to implement the mitigation measures required by the local government, as they may be quite costly.

The importance of the Environmental Impact Assessment as an effective tool for the purpose of integrating environmental considerations with development planning is highly recognized in Sri Lanka. The application of this technique is considered as a means of ensuring that the likely effects of new development projects on the environment are fully understood and taken into account before development is allowed to proceed. The importance of this management tool to foresee potential environmental impacts and problems caused by proposed projects and its use as a mean to make project more suitable to the environment are highly appreciated.

## United States

The National Environmental Policy Act of 1969 (NEPA), enacted in 1970, established a policy of environmental impact assessment for federal agency actions, federally funded activities or federally permitted/licensed activities that in the U. S. is termed "environmental review" or simply "the NEPA process." The law also created the Council on Environmental Quality, which promulgated regulations to codify the law's requirements. Under United States environmental law an Environmental Assessment (EA) is compiled to determine the need for an *Environmental Impact Statement* (EIS). Federal or federalized actions expected to subject or be subject to significant environmental impacts will publish a Notice of Intent to Prepare an EIS as soon as significance is known. Certain actions of federal agencies must be preceded by the NEPA process. Contrary to a widespread misconception, NEPA does not prohibit the federal government or its licensees/permittees from harming the environment, nor does it specify any penalty if an environmental impact assessment turns out to be inaccurate, intentionally or otherwise. NEPA requires that plausible statements as to the prospective impacts be disclosed in advance. The purpose of NEPA process is to ensure that the decision maker is fully informed of the environmental aspects and consequences prior to making the final decision.

## Environmental Assessment

An environmental assessment (EA) is an environmental analysis prepared pursuant to the National Environmental Policy Act to determine whether a federal action would significantly affect the environment and thus require a more detailed *Environmental Impact Statement (EIS).* The certified release of an Environmental Assessment results in either a *Finding of No Significant Impact (FONSI)* or an EIS.

The Council on Environmental Quality (CEQ), which oversees the administration of NEPA, issued regulations for implementing the NEPA in 1979. Eccleston reports that the NEPA regulations barely mention preparation of EAs. This is because the EA was originally intended to be a simple document used in relatively rare instances where an agency was not sure if the potential significance of an action would be sufficient to trigger preparation of an EIS. But today, because EISs are so much longer and complicated to prepare, federal agencies are going to great effort to avoid preparing EISs by using EAs, even in cases where the use of EAs may be inappropriate. The ratio of EAs that are being issued compared to EISs is about 100 to 1.

Likewise, even the preparation of an accurate EA is viewed today as an onerous burden by many entities responsible for the environmental review of a proposal. Federal agencies have responded by streamlining their regulations that implement NEPA environmental review, by defining categories of projects that by their well understood nature may be safely excluded from review under NEPA, and by drawing up lists of project types that have negligible material impact upon the environment and can thus be exempted.

## Content

The Environmental Assessment is a concise public document prepared by the federal action agency that serves to:

1. briefly provide sufficient evidence and analysis for determining whether to prepare an EIS or a Finding of No Significant Impact (FONSI)
2. Demonstrate compliance with the act when no EIS is required
3. facilitate the preparation of an EIS when a FONSI cannot be demonstrated

The Environmental Assessment includes a brief discussion of the purpose and need of the proposal and of its alternatives as required by NEPA 102(2)(E), and of the human environmental impacts resulting from and occurring to the proposed actions and alternatives considered practicable, plus a listing of studies conducted and agencies and stakeholders consulted to reach these conclusions. The action agency must approve an EA before it is made available to the public. The EA is made public through notices of availability by local, state, or regional clearing houses, often triggered by the purchase of a public notice advertisement in a newspaper of general circulation in the proposed activity area.

## Structure

The structure of a generic Environmental Assessment is as follows:

1. Summary
2. Introduction
    - Background

- Purpose and Need for Action
- Proposed Action
- Decision Framework
- Public Involvement
- Issues

3. Alternatives, including the Proposed Action
    - Alternatives
    - Mitigation Common to All Alternatives
    - Comparison of Alternatives
4. Environmental Consequences
5. Consultation and Coordination

## Procedure

The EA becomes a draft public document when notice of it is published, usually in a newspaper of general circulation in the area affected by the proposal. There is a 15-day review period required for an Environmental Assessment (30 days if exceptional circumstances) while the document is made available for public commentary, and a similar time for any objection to improper process. Commenting on the Draft EA is typically done in writing or email, submitted to the lead action agency as published in the notice of availability. An EA does not require a public hearing for verbal comments. Following the mandated public comment period, the lead action agency responds to any comments, and certifies either a FONSI or a Notice of Intent (NOI) to prepare an EIS in its public environmental review record. The preparation of an EIS then generates a similar but more lengthy, involved and expensive process.

## Environmental Impact Statement

The adequacy of an environmental impact statement (EIS) can be challenged in federal court. Major proposed projects have been blocked because of an agency's failure to prepare an acceptable EIS. One prominent example was the Westway landfill and highway development in and along the Hudson River in New York City. Another prominent case involved the Sierra Club suing the Nevada Department of Transportation over its denial of the club's request to issue a supplemental EIS addressing air emissions of particulate matter and hazardous air pollutants in the case of widening U.S. Route 95 through

Las Vegas. The case reached the United States Court of Appeals for the Ninth Circuit, which led to construction on the highway being halted until the court's final decision. The case was settled prior to the court's final decision.

Several state governments that have adopted "little NEPAs," state laws imposing EIS requirements for particular state actions. Some those state laws such as the California Environmental Quality Act refer to the required environmental impact study as an environmental impact report.

These variety of state requirements are yielding voluminous data not just upon impacts of individual projects, but also to elucidate scientific areas that had not been sufficiently researched. For example, in a seemingly routine *Environmental Impact Report* for the city of Monterey, California, information came to light that led to the official federal endangered species listing of Hickman's potentilla, a rare coastal wildflower.

## Transboundary Application

Environmental threats do not respect national borders. International pollution can have detrimental effects on the atmosphere, oceans, rivers, aquifers, farmland, the weather and biodiversity. Global climate change is transnational. Specific pollution threats include acid rain, radioactive contamination, debris in outer space, stratospheric ozone depletion and toxic oil spills. The Chernobyl disaster, precipitated by a nuclear accident on April 26, 1986, is a stark reminder of the devastating effects of transboundary nuclear pollution.

Environmental protection is inherently a cross-border issue and has led to the creation of transnational regulation via multilateral and bilateral treaties. The United Nations Conference on the Human Environment (UNCHE or Stockholm Conference) held in Stockholm in 1972 and the United Nations Conference on the Environment and Development (UNCED or Rio Summit, Rio Conference, or Earth Summit) held in Rio de Janeiro in 1992 were key in the creation of about 1,000 international instruments that include at least some provisions related to the environment and its protection.

The United Nations Economic Commission for Europe's Convention on Environmental Impact Assessment in a Transboundary Context was negotiated to provide an international legal framework for transboundary EIA.

However, as there is no universal legislature or administration with a comprehensive mandate, most international treaties exist parallel to one another and are further developed without the benefit of consideration being given to potential conflicts with other agreements. There is also the issue of international enforcement. This has led to duplications and failures, in part due to an inability to enforce agreements. An example is the failure of many international fisheries regimes to restrict harvesting practises.

## Criticism

As per Jay *et al.*, EIA is used as a decision aiding tool rather than decision making tool. There is growing dissent about them as their influence on decisions is limited. Improved training for practitioners, guidance on best practice and continuing research have all been proposed.

EIAs have been criticized for excessively limiting their scope in space and time. No accepted procedure exists for determining such boundaries. The boundary refers to 'the spatial and temporal boundary of the proposal's effects'. This boundary is determined by the applicant and the lead assessor, but in practice, almost all EIAs address only direct and immediate on-site effects.

Development causes both direct and indirect effects. Consumption of goods and services, production, use and disposal of building materials and machinery, additional land use for activities of manufacturing and services, mining and refining, etc., all have environmental impacts. The indirect effects of development can be much higher than the direct effects examined by an EIA. Proposals such as airports or shipyards cause wide-ranging national and international effects, which should be covered in EIAs.

Broadening the scope of EIA can benefit the conservation of threatened species. Instead of concentrating on the project site, some EIAs employed a habitat-based approach that focused on much broader relationships among humans and the environment. As a result, alternatives that reduce the negative effects to the population of whole species, rather than local subpopulations, can be assessed.

Thissen and Agusdinata have argued that little attention is given to the systematic identification and assessment of uncertainties in environmental studies which is critical in situations where uncertainty cannot be easily reduced by doing more research. In line with this, Maier et al. have concluded on the need to consider uncertainty at all stages of the decision-making process. In such a way decisions can be made with confidence or known uncertainty. These proposals are justified on data that shows that environmental assessments fail to predict accurately the impacts observed. Tenney et al. and Wood et al. have reported evidence of the intrinsic uncertainty attached to EIAs predictions from a number of case studies worldwide. The gathered evidence consisted of comparisons between predictions in EIAs and the impacts measured during, or following project implementation. In explaining this trend, Tenney et al. have highlighted major causes such as project changes, modelling errors, errors in data and assumptions taken and bias introduced by people in the projects analyzed. Cardenas and Halman provide a comprehensive review on the issues of uncertainty in environmental impact assessments.

## Public Health

Public health refers to "the science and art of preventing disease, prolonging life and promoting health through organized efforts and informed choices of society, organizations, public and private, communities and individuals." It is concerned with threats to health based on population health analysis. The population in question can be as small as a handful of people, or as large as all the inhabitants of several continents (for instance, in the case of a pandemic). The dimensions of health can encompass "a state of complete physical, mental and social well-being and not merely the absence of disease or infirmity", as defined by the United Nations' World Health Organization. Public health incorporates the interdisciplinary approaches of epidemiology, biostatistics and health services. Environmental health, community health, behavioral health, health economics, public policy, insurance medicine, mental health and occupational safety and health are other important subfields.

Newspaper headlines from around the world about polio vaccine tests (13 April 1955)

The focus of public health intervention is to improve health and quality of life through prevention and treatment of disease and other physical and mental health conditions. This is done through surveillance of cases and health indicators, and through promotion of healthy behaviors. Examples of common public health measures include promotion of hand washing, breastfeeding, delivery of vaccinations, and distribution of condoms to control the spread of sexually transmitted diseases.

Modern public health practice requires multidisciplinary teams of public health workers and professionals including physicians specializing in public health/community medicine/infectious disease, psychologists, epidemiologists, biostatisticians, medical assistants or Assistant Medical Officers, public health nurses, midwives, medical microbiologists, environmental health officers / public health inspectors, pharmacists, dentists, dietitians and nutritionists, veterinarians, public health engineers, public health lawyers, sociologists, community development workers, communications experts, bioethicists, and others.

There is a great disparity in access to health care and public health initiatives between developed nations and developing nations. In the developing world, public health infrastructures are still forming.

## Background

The focus of a public health intervention is to prevent and manage diseases, injuries and other health conditions through surveillance of cases and the promotion of healthy behaviors, communities and environments. Many diseases are preventable through simple, non-medical methods. For example, research has shown that the simple act of hand washing with soap can prevent many contagious diseases. In other cases, treating a disease or controlling a pathogen can be vital to preventing its spread to others, such as during an outbreak of infectious disease, or contamination of food or water supplies. Public health communications programs, vaccination programs, and distribution of condoms are examples of common public health measures. Measures such as these have contributed greatly to the health of populations and increases in life expectancy.

Public health plays an important role in disease prevention efforts in both the developing world and in developed countries, through local health systems and non-governmental organizations. The World Health Organization (WHO) is the international agency that coordinates and acts on global public health issues. Most countries have their own government public health agencies, sometimes known as ministries of health, to respond to domestic health issues. For example, in the United States, the front line of public health initiatives are state and local health departments. The United States Public Health Service (PHS), led by the Surgeon General of the United States, and the Centers for Disease Control and Prevention, headquartered in Atlanta, are involved with several international health activities, in addition to their national duties. In Canada, the Public Health Agency of Canada is the national agency responsible for public health, emergency preparedness and response, and infectious and chronic disease control and prevention. The Public health system in India is managed by the Ministry of Health & Family Welfare of the government of India with state owned health care facilities.

## Current Practice

## Public Health Programs

Most governments recognize the importance of public health programs in reducing the incidence of disease, disability, and the effects of aging and other physical and mental health conditions, although public health generally receives significantly less government funding compared with medicine. Public health programs providing vaccinations have made strides in promoting health, including the eradication of smallpox, a disease that plagued humanity for thousands of years.

Three former directors of the Global Smallpox Eradication Programme read the news that smallpox had been globally eradicated, 1980

The World Health Organization (WHO) identifies core functions of public health programs including:

- providing leadership on matters critical to health and engaging in partnerships where joint action is needed;
- shaping a research agenda and stimulating the generation, translation and dissemination of valuable knowledge;
- setting norms and standards and promoting and monitoring their implementation;
- articulating ethical and evidence-based policy options;
- monitoring the health situation and assessing health trends.

In particular, public health surveillance programs can:

- serve as an early warning system for impending public health emergencies;
- document the impact of an intervention, or track progress towards specified goals; and
- monitor and clarify the epidemiology of health problems, allow priorities to be set, and inform health policy and strategies.
- diagnose, investigate, and monitor health problems and health hazards of the community

Public health surveillance has led to the identification and prioritization of many public health issues facing the world today, including HIV/AIDS, diabetes, waterborne diseases, zoonotic diseases, and antibiotic resistance leading to the reemergence of infectious diseases such as tuberculosis. Antibiotic resistance, also known as drug resistance, was the theme of World Health Day 2011. Although the prioritization of pressing public health issues is important, Laurie Garrett argues that there are following consequences. When foreign aid is funneled into disease-specific programs, the importance of public health in general is disregarded. This public health problem of stovepiping is thought to create a lack of funds to combat other existing diseases in a given country.

For example, the WHO reports that at least 220 million people worldwide suffer from diabetes. Its incidence is increasing rapidly, and it is projected that the number of diabetes deaths will double by the year 2030. In a June 2010 editorial in the medical journal *The Lancet*, the authors opined that "The fact that type 2 diabetes, a largely preventable disorder, has reached epidemic proportion is a public health humiliation." The risk of type 2 diabetes is closely linked with the growing problem of obesity. The WHO's latest estimates highlighted that globally approximately 1.5 billion adults were overweight in 2008, and nearly 43 million children under the age of five were overweight in 2010. The United States is the leading country with 30.6% of its population being obese. Mexico follows behind with 24.2% and the United Kingdom with 23%. Once considered a problem in high-income countries, it is now on the rise in low-income countries, especially in urban settings. Many public health programs are increasingly dedicating attention and resources to the issue of obesity, with objectives to address the underlying causes including healthy diet and physical exercise.

Some programs and policies associated with public health promotion and prevention can be controversial. One such example is programs focusing on the prevention of HIV transmission through safe sex campaigns and needle-exchange programmes. Another is the control of tobacco smoking. Changing smoking behavior requires long-term strategies, unlike the fight against communicable diseases, which usually takes a shorter period for effects to be observed. Many nations have implemented major initiatives to cut smoking, such as increased taxation and bans on smoking in some or all public places. Proponents argue by presenting evidence that smoking is one of the major killers, and that therefore governments have a duty to reduce the death rate, both through limiting passive (second-hand) smoking and by providing fewer opportunities for people to smoke. Opponents say that this undermines individual freedom and personal responsibility, and worry that the state may be emboldened to remove more and more choice in the name of better population health overall.

Simultaneously, while communicable diseases have historically ranged uppermost as a global health priority, non-communicable diseases and the underlying behavior-related risk factors have been at the bottom. This is changing however, as illustrated by the United Nations hosting its first General Assembly Special Summit on the issue of non-communicable diseases in September 2011.

Many health problems are due to maladaptive personal behaviors. From an evolutionary psychology perspective, over consumption of novel substances that are harmful is due to the activation of an evolved reward system for substances such as drugs, tobacco, alcohol, refined salt, fat, and carbohydrates. New technologies such as modern transportation also cause reduced physical activity. Research has found that behavior is more effectively changed by taking evolutionary motivations into consideration instead of only presenting information about health effects. Thus, the increased use of soap and hand-washing to prevent diarrhea is much more effectively promoted if its lack of use is associated with the emotion of disgust. Disgust is an evolved system for avoiding contact with substances that spread infectious diseases. Examples might include films that show how fecal matter contaminates food. The marketing industry has long known the importance of associating products with high status and attractiveness to others. Conversely, it has been argued that emphasizing the harmful and undesirable effects of tobacco smoking on other persons and imposing smoking bans in public places have been particularly effective in reducing tobacco smoking.

## Applications in Health Care

As well as seeking to improve population health through the implementation of specific population-level interventions, public health contributes to medical care by identifying and assessing population needs for health care services, including:

- Assessing current services and evaluating whether they are meeting the objectives of the health care system
- Ascertaining requirements as expressed by health professionals, the public and other stakeholders
- Identifying the most appropriate interventions
- Considering the effect on resources for proposed interventions and assessing their cost-effectiveness
- Supporting decision making in health care and planning health services including any necessary changes.
- Informing, educating, and empowering people about health issues

## Implementing Effective Improvement Strategies

To improve public health, one important strategy is to promote modern medicine and scientific neutrality to drive the public health policy and campaign, which is recommended by Armanda Solorzana, through a case study of the Rockefeller Foundation's hookworm campaign in Mexico in the 1920s. Soloranza argues that public health policy can't concern only politics or economics. Political concerns can lead government officials to hide the real numbers of people affected by disease in their regions, such as upcoming elections. Therefore, scientific neutrality in making public health policy is critical; it can ensure treatment needs are met regardless of political and economic conditions.

The history of public health care clearly shows the global effort to improve health care for all. However, in modern-day medicine, real, measurable change has not been clearly seen, and critics argue that this lack of improvement is due to ineffective methods that are being implemented. As argued by Paul E. Farmer, structural interventions could possibly have a large impact, and yet there are numerous problems as to why this strategy has yet to be incorporated into the health system. One of the main reasons that he suggests could be the fact that physicians are not properly trained to carry out structural interventions, meaning that the ground level health care professionals cannot implement these improvements. While structural interventions can not be the only area for improvement, the lack of coordination between socioeconomic factors and health care for the poor could be counterproductive, and end up causing greater inequity between the health care services received by the rich and by the poor. Unless health care is no longer treated as a commodity, global public health can ultimately not be achieved. This being the case, without changing the way in which health care is delivered to those who have less access to it, the universal goal of public health care cannot be achieved.

## Public Health 2.0

*Public Health 2.0* is a movement within public health that aims to make the field more accessible to the general public and more user-driven. The term is used in three senses. In the first sense, "Public Health 2.0" is similar to "Health 2.0" and describes the ways in which traditional public health practitioners and institutions are reaching out (or could reach out) to the public through social media and health blogs.

In the second sense, "Public Health 2.0" describes public health research that uses data gathered from social networking sites, search engine queries, cell phones, or other technologies. A recent example is the proposal of statistical framework that utilizes online user-generated content (from social media or search engine queries) to estimate the impact of an influenza vaccination campaign in the UK.

In the third sense, "Public Health 2.0" is used to describe public health activities that are completely user-driven. An example is the collection and sharing of information about environmental radiation levels after the March 2011 tsunami in Japan. In all cases, Public Health 2.0 draws on ideas from Web 2.0, such as crowdsourcing, information sharing, and user-centred design. While many individual healthcare providers have started making their own personal contributions to "Public Health 2.0" through personal blogs, social profiles, and websites, other larger organizations, such as the American Heart Association (AHA) and United Medical Education (UME), have a larger team of employees centered around online driven health education, research, and training. These private organizations recognize the need for free and easy to access health materials often building libraries of educational articles.

## Developing Countries

There is a great disparity in access to health care and public health initiatives between developed nations and developing nations. In the developing world, public health infrastructures are still forming. There may not be enough trained health workers or monetary resources to provide even a basic level of medical care and disease prevention. As a result, a large majority of disease and mortality in the developing world results from and contributes to extreme poverty. For example, many African governments spend less than US$10 per person per year on health care, while, in the United States, the federal government spent approximately US$4,500 per capita in 2000. However, expenditures on health care should not be confused with spending on public health. Public health measures may not generally be considered "health care" in the strictest sense. For example, mandating the use of seat belts in cars can save countless lives and contribute to the health of a population, but typically money spent enforcing this rule would not count as money spent on health care.

Emergency Response Team in Burma after Cyclone Nargis in 2008

Large parts of the developing world remained plagued by largely preventable or treatable infectious diseases and poor maternal and child health, exacerbated by malnutrition and poverty. The WHO reports that a lack of exclusive breastfeeding during the first six months of life contributes to over a million avoidable child deaths each year. Intermittent preventive therapy aimed at treating and preventing malaria episodes among pregnant women and young children is one public health measure in endemic countries.

Each day brings new front-page headlines about public health: emerging infectious diseases such as SARS, rapidly making its way from China to Canada, the United States and other geographically distant countries; reducing inequities in health care access through publicly funded health insurance programs; the HIV/AIDS pandemic and its spread from certain high-risk groups to the general population in many countries, such as in South Africa; the increase of childhood obesity and the concomitant increase in type II diabetes among children; the social, economic and health effects of adolescent pregnancy; and the public health challenges related to natural disasters such as the 2004 Indian Ocean tsunami, 2005's Hurricane Katrina in the United States and the 2010 Haiti earthquake.

Since the 1980s, the growing field of population health has broadened the focus of public health from individual behaviors and risk factors to population-level issues such as inequality, poverty, and education. Modern public health is often concerned with addressing determinants of health across a population. There is a recognition that our health is affected by many factors including where we live, genetics, our income, our educational status and our social relationships; these are known as "social determinants of health." A social gradient in health runs through society. The poorest generally suffer the worst health, but even the middle classes will generally have worse health outcomes than those of a higher social stratum. The new public health advocates for population-based policies that improve health in an equitable manner.

## Sustainable Development Goals

To address current and future challenges in addressing health issues in the world, the United Nations have developed the Sustainable Development Goals 2015 building off of the Millennium Development Goals of 2000 to be completed by 2030. These goals in their entirety encompass the entire spectrum of development across nations, however Goals 1-6 directly address health disparities, primarily in developing countries. These six goals address key issues in global public health: Poverty, Hunger and food security, Health, Education, Gender equality and women's empowerment, and water and sanitation. Public health officials can use these goals to set their own agenda and plan for smaller scale initiatives for their organizations. These goals hope to lessen the burden of disease and inequality faced by developing countries and lead to a healthier future.

The links between the various sustainable development goals and public health are numerous and well established:

- Living below the poverty line is attributed to poorer health outcomes and can be even worse for persons living in developing countries where extreme poverty is more common. A child born into poverty is twice as likely to die before the age of five compared to a child from a wealthier family.
- The detrimental effects of hunger and malnutrition that can arise from systemic challenges with food security are enormous. The World Health Organization estimates that 12.9 percent of the population in developing countries is undernourished.
- Health challenges in the developing world are enormous, with "only half of the women in developing nations receiving the recommended amount of healthcare they need.

- Educational equity has yet to be reached in the world. Public health efforts are impeded by this, as a lack of education can lead to poorer health outcomes. This is shown by children of mothers who have no education having a lower survival rate compared to children born to mothers with primary or greater levels of education. Cultural differences in the role of women vary by country, many gender inequalities are found in developing nations. Combating these inequalities has shown to also lead to better public health outcome.
- In studies done by the World Bank on populations in developing countries, it was found that when women had more control over household resources, the children benefit through better access to food, healthcare, and education.
- Basic sanitation resources and access to clean sources of water are a basic human right. However, 1.8 billion people globally use a source of drinking water that is fecally contaminated, and 2.4 billion people lack access to basic sanitation facilities like toilets or pit latrines. A lack of these resources is what causes approximately 1000 children a day to die from diarrhoel diseases that could have been prevented from better water and sanitation infrastructure.

## Education and Training

Education and training of public health professionals is available throughout the world in Schools of Public Health, Medical Schools, Veterinary Schools, Schools of Nursing, and Schools of Public Affairs. The training typically requires a university degree with a focus on core disciplines of biostatistics, epidemiology, health services administration, health policy, health education, behavioral science and environmental health. In the global context, the field of public health education has evolved enormously in recent decades, supported by institutions such as the World Health Organization and the World Bank, among others. Operational structures are formulated by strategic principles, with educational and career pathways guided by competency frameworks, all requiring modulation according to local, national and global realities. It is critically important for the health of populations that nations assess their public health human resource needs and develop their ability to deliver this capacity, and not depend on other countries to supply it.

## Schools of Public Health - A U.S. Perspective

In the United States, the Welch-Rose Report of 1915 has been viewed as the basis for the critical movement in the history of the institutional schism between public health and medicine because it led to the establishment of schools of public health supported by the Rockefeller Foundation. The report was authored by William Welch, founding dean of the Johns Hopkins Bloomberg School of Public Health, and Wickliffe Rose of the Rockefeller Foundation. The report focused more on research than practical education. Some have blamed the Rockefeller Foundation's 1916 decision to support the establishment of schools of public health for creating the schism between public health and medicine and legitimizing the rift between medicine's laboratory investigation of the mechanisms of disease and public health's nonclinical concern with environmental and social influences on health and wellness.

Even though schools of public health had already been established in Canada, Europe and North Africa, the United States had still maintained the traditional system of housing faculties of public health within their medical institutions. A $25,000 donation from businessman Samuel Zemurray instituted the School of Public Health and Tropical Medicine at Tulane University in 1912 conferring its first doctor of public health degree in 1914. The Johns Hopkins School of Hygiene and Public Health became an independent, degree-granting institution for research and training in public health, and the largest public health training facility in the United States, when it was founded in 1916. By 1922, schools of public health were established at Columbia, Harvard and Yale on the Hopkins model. By 1999 there were twenty nine schools of public health in the US, enrolling around fifteen thousand students.

Over the years, the types of students and training provided have also changed. In the beginning, students who enrolled in public health schools typically had already obtained a medical degree; public health school training was largely a second degree for medical professionals. However, in 1978, 69% of American students enrolled in public health schools had only a bachelor's degree.

## Degrees in Public Health

Schools of public health offer a variety of degrees which generally fall into two categories: professional or academic. The two major postgraduate degrees are the Master of Public Health (M.P.H.) or the Master of Science in Public Health (MSPH). Doctoral studies in this field include Doctor of Public Health (DrPH) and Doctor of Philosophy (Ph.D.) in a subspeciality of greater Public Health disciplines. DrPH is regarded as a professional leadership degree and Ph.D. as more of an academic degree.

Professional degrees are oriented towards practice in public health settings. The Master of Public Health, Doctor of Public Health, Doctor of Health Science (DHSc) and the Master of Health Care Administration are examples of degrees which are geared towards people who want careers as practitioners of public health in health departments, managed care and community-based organizations, hospitals and consulting firms among others. Master of Public Health degrees broadly fall into two categories, those that put more emphasis on an understanding of epidemiology and statistics as the scientific basis of public health practice and those that include a more eclectic range of methodologies. A Master of Science of Public Health is similar to an MPH but is considered an academic degree (as opposed to a professional degree) and places more emphasis on scientific methods and research. The same distinction can be made between the DrPH and the DHSc. The DrPH is considered a professional degree and the DHSc is an academic degree.

Academic degrees are more oriented towards those with interests in the scientific basis of public health and preventive medicine who wish to pursue careers in research, university teaching in graduate programs, policy analysis and development, and other high-level public health positions. Examples of academic degrees are the Master of Science, Doctor of Philosophy, Doctor of Science (ScD), and Doctor of Health Science (DHSc). The doctoral programs are distinct from the MPH and other professional programs by the addition of advanced coursework and the nature and scope of a dissertation research project.

In the United States, the Association of Schools of Public Health represents Council on Education for Public Health (CEPH) accredited schools of public health. Delta Omega is the honor society for graduate studies in public health. The society was founded in 1924 at the Johns Hopkins School of Hygiene and Public Health. Currently, there are approximately 68 chapters throughout the United States and Puerto Rico.

## History

### Early History

Public health has early roots in antiquity. From the beginnings of human civilization, it was recognized that polluted water and lack of proper waste disposal spread communicable diseases (theory of miasma). Early religions attempted to regulate behavior that specifically related to health, from types of food eaten, to regulating certain indulgent behaviors, such as drinking alcohol or sexual relations. Leaders were responsible for the health of their subjects to ensure social stability, prosperity, and maintain order.

The primitive nature of medieval medicine rendered Europe helpless to the onslaught of the Black Death in the 14th century. Fragment of a miniature from "The Chronicles of Gilles Li Muisis" (1272-1352). Bibliothèque royale de Belgique, MS 13076-77, f. 24v.

By Roman times, it was well understood that proper diversion of human waste was a necessary tenet of public health in urban areas. The ancient Chinese medical doctors developed the practice of variolation following a smallpox epidemic around 1000 BC. An individual without the disease could gain some measure of immunity against it by inhaling the dried crusts that formed around lesions of infected individuals. Also, children were protected by inoculating a scratch on their forearms with the pus from a lesion.

In 1485 the Republic of Venice established a Permanent Court of supervisors of health with special attention to the prevention of the spread of epidemics in the territory from abroad. The three supervisors were initially appointed by the Venetian Senate. In 1537 it was assumed by the Grand Council, and in 1556 added two judges, with the task of control, on behalf of the Republic, the efforts of the supervisors.

However, according to Michel Foucault, the plague model of governmentality was later controverted by the cholera model. A Cholera pandemic devastated Europe between 1829 and 1851, and was first fought by the use of what Foucault called "social medicine", which focused on flux, circulation of air, location of cemeteries, etc. All those concerns, born of the miasma theory of disease, were mixed with urbanistic concerns for the management of populations, which Foucault designated as the concept of "biopower". The German conceptualized this in the *Polizeiwissenschaft* ("Police science").

## Modern Public Health

The 18th century saw rapid growth in voluntary hospitals in England. The latter part of the century brought the establishment of the basic pattern of improvements in public health over the next two centuries: a social evil was identified, private philanthropists brought attention to it, and changing public opinion led to government action.

1802 caricature of Edward Jenner vaccinating patients who feared it would make them sprout cowlike appendages.

The practice of vaccination became prevalent in the 1800s, following the pioneering work of Edward Jenner in treating smallpox. James Lind's discovery of the causes of scurvy amongst sailors and its mitigation via the introduction of fruit on lengthy voyages was published in 1754 and led to the adoption of this idea by the Royal Navy. Efforts were also made to promulgate health matters to the broader public; in

1752 the British physician Sir John Pringle published *Observations on the Diseases of the Army in Camp and Garrison*, in which he advocated for the importance of adequate ventilation in the military barracks and the provision of latrines for the soldiers.

With the onset of the Industrial Revolution, living standards amongst the working population began to worsen, with cramped and unsanitary urban conditions. In the first four decades of the 19th century alone, London's population doubled and even greater growth rates were recorded in the new industrial towns, such as Leeds and Manchester. This rapid urbanisation exacerbated the spread of disease in the large conurbations that built up around the workhouses and factories. These settlements were cramped and primitive with no organized sanitation. Disease was inevitable and its incubation in these areas was encouraged by the poor lifestyle of the inhabitants. Unavailable housing led to the rapid growth of slums and the per capita death rate began to rise alarmingly, almost doubling in Birmingham and Liverpool. Thomas Malthus warned of the dangers of overpopulation in 1798. His ideas, as well as those of Jeremy Bentham, became very influential in government circles in the early years of the 19th century.

## Public Health Legislation

The first attempts at sanitary reform and the establishment of public health institutions were made in the 1840s. Thomas Southwood Smith, physician at the London Fever Hospital, began to write papers on the importance of public health, and was one of the first physicians brought in to give evidence before the Poor Law Commission in the 1830s, along with Neil Arnott and James Phillips Kay. Smith advised the government on the importance of quarantine and sanitary improvement for limiting the spread of infectious diseases such as cholera and yellow fever.

Sir Edwin Chadwick was a pivotal influence on the early public health campaign.

The Poor Law Commission reported in 1838 that "the expenditures necessary to the adoption and maintenance of measures of prevention would ultimately amount to less than the cost of the disease now constantly engendered". It recommended the implementation of large scale government engineering projects to alleviate the conditions that allowed for the propagation of disease. The Health of Towns Association was formed in Exeter on 11 December 1844, and vigorously campaigned for the development of public health in the United Kingdom. Its formation followed the 1843 establishment of the Health of Towns Commission, chaired by Sir Edwin Chadwick, which produced a series of reports on poor and insanitary conditions in British cities.

These national and local movements led to the Public Health Act, finally passed in 1848. It aimed to improve the sanitary condition of towns and populous places in England and Wales by placing the supply of water, sewerage, drainage, cleansing and paving under a single local body with the General Board of Health as a central authority. The Act was passed by the Liberal government of Lord John Russell, in response to the urging of Edwin Chadwick. Chadwick's seminal report on *The Sanitary Condition of the Labouring Population* was published in 1842 and was followed up with a supplementary report a year later.

Vaccination for various diseases was made compulsory in the United Kingdom in 1851, and by 1871 legislation required a comprehensive system of registration run by appointed vaccination officers.

Further interventions were made by a series of subsequent Public Health Acts, notably the 1875 Act. Reforms included latrinization, the building of sewers, the regular collection of garbage followed by incineration or disposal in a landfill, the provision of clean water and the draining of standing water to prevent the breeding of mosquitoes.

The Infectious Disease (Notification) Act 1889 mandated the reporting of infectious diseases to the local sanitary authority, which could then pursue measures such as the removal of the patient to hospital and the disinfection of homes and properties.

In the U.S., the first public health organization based on a state health department and local boards of health was founded in New York City in 1866.

## Epidemiology

The science of epidemiology was founded by John Snow's identification of a polluted public water well as the source of an 1854 cholera outbreak in London. Dr. Snow believed in the germ theory of disease as opposed to the prevailing miasma theory. He first publicized his theory in an essay, *On the Mode of Communication of Cholera*, in 1849, followed by a more detailed treatise in 1855 incorporating the results of his investigation of the role of the water supply in the Soho epidemic of 1854.

By talking to local residents (with the help of Reverend Henry Whitehead), he identified the source of the outbreak as the public water pump on Broad Street (now Broadwick Street). Although Snow's chemical and microscope examination of a water sample from the Broad Street pump did not conclusively prove its danger, his studies of the pattern of the disease were convincing enough to persuade the local council to disable the well pump by removing its handle.

John Snow's dot map, showing the clusters of cholera cases in the London epidemic of 1854.

Snow later used a dot map to illustrate the cluster of cholera cases around the pump. He also used statistics to illustrate the connection between the quality of the water source and cholera cases. He showed that the Southwark and Vauxhall Waterworks Company was taking water from sewage-polluted sections of the Thames and delivering the water to homes, leading to an increased incidence of cholera. Snow's study was a major event in the history of public health and geography. It is regarded as the founding event of the science of epidemiology.

## Disease Control

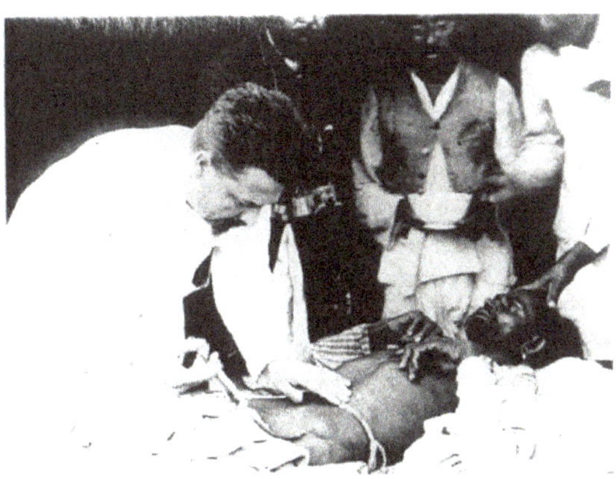

Paul-Louis Simond injecting a plague vaccine in Karachi, 1898.

With the pioneering work in bacteriology of French chemist Louis Pasteur and German scientist Robert Koch, methods for isolating the bacteria responsible for a given disease and vaccines for remedy were developed at the turn of the 20th century. British physician Ronald Ross identified the mosquito as the carrier of malaria and laid the foundations for combating the disease. Joseph Lister revolutionized surgery by the introduction of antiseptic surgery to eliminate infection. French epidemiologist Paul-Louis Simond proved that plague was carried by fleas on the back of rats, and the Americans Walter Reed and James Carroll demonstrated that mosquitoes carry the virus responsible for yellow fever.

With onset of the epidemiological transition and as the prevalence of infectious diseases decreased through the 20th century, public health began to put more focus on chronic diseases such as cancer and heart disease. Previous efforts in many developed countries had already led to dramatic reductions in the infant mortality rate using preventative methods. In Britain, the infant mortality rate fell from over 15% in 1870 to 7% by 1930.

## United States

In the administration of the second president of the United States John Adams, the Congress authorized the creation of hospitals for mariners. As the U.S. expanded, the scope of the governmental health agency expanded.

Seal of the United States Public Health Service

In the United States, public health worker Sara Josephine Baker, M.D. established many programs to help the poor in New York City keep their infants healthy, leading teams of nurses into the crowded neighborhoods of Hell's Kitchen and teaching mothers how to dress, feed, and bathe their babies.

Public health nursing made available through child welfare services in U.S. (c. 1930s)

Another key pioneer of public health in the U.S. was Lillian Wald, who founded the Henry Street Settlement house in New York. The Visiting Nurse Service of New York was a significant organization for bringing health care to the urban poor.

Logo of the Visiting Nurse Service of New York, founded by nursing pioneer Lillian Wald in 1893

Dramatic increases in average life span in the late 19th century and 20th century, is widely credited to public health achievements, such as vaccination programs and control of many infectious diseases including polio, diphtheria, yellow fever and smallpox; effective health and safety policies such as road traffic safety and occupational safety; improved family planning; tobacco control measures; and programs designed to decrease non-communicable diseases by acting on known risk factors such as a person's background, lifestyle and environment.

Another major public health improvement was the decline in the "urban penalty" brought about by improvements in sanitation. These improvements included chlorination of drinking water, filtration and sewage treatment which led to the decline in deaths caused by infectious waterborne diseases such as cholera and intestinal diseases. The federal Office of Indian Affairs (OIA) operated a large-scale field nursing program. Field nurses targeted native women for health education, emphasizing personal hygiene and infant care and nutrition.

## Latin America

Public health issues were important for the Spanish empire during the colonial era. Epidemic disease was the main factor in the decline of indigenous populations in the era immediately following the sixteenth-century conquest era and was a problem during the colonial era. The Spanish crown took steps in eighteenth-century Mexico to bring in regulations to make populations healthier. In the late nineteenth century, Mexico was in the process of modernization, and public health issues were again tackled from a scientific point of view. Even during the Mexican Revolution (1910–20), public health was an important concern, with a text on hygiene published in 1916. During the Mexican Revolution, feminist and trained nurse Elena Arizmendi Mejia founded the Neutral White Cross, treating wounded soldiers no matter for what faction they fought. In the post-revolutionary period after 1920, improved public health was a revolutionary goal of the Mexican government.

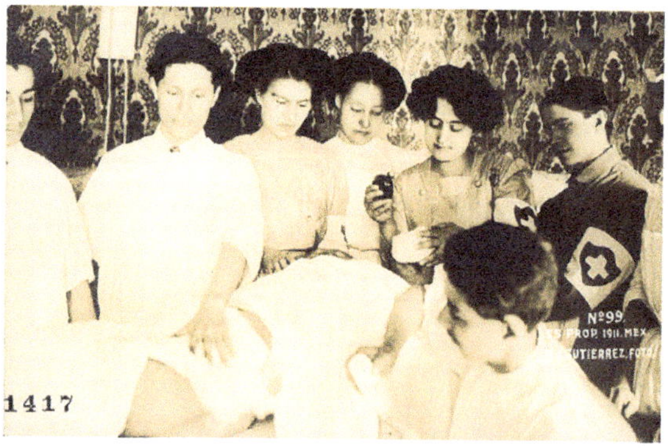


Elena Arizmendi Mejia and volunteers of the Mexican Neutral White Cross, 1911

Public health was important elsewhere in Latin America in consolidating state power and integrating marginalized populations into the nation-state. In Colombia, public health was a means for creating and implementing ideas of citizenship. In Bolivia, the push came after their 1952 revolution.

# Industrial Wastewater Treatment

Industrial wastewater treatment covers the mechanisms and processes used to treat wastewater that is produced as a by-product of industrial or commercial activities. After treatment, the treated industrial wastewater (or effluent) may be reused or released to a sanitary sewer or to a surface water in the environment. Most industries produce some wastewater although recent trends in the developed world have been to minimise such production or recycle such wastewater within the production process. However, many industries remain dependent on processes that produce wastewaters.

## Sources of Industrial Wastewater

## Complex Organic Chemicals Industry

A range of industries manufacture or use complex organic chemicals. These include pesticides, pharmaceuticals, paints and dyes, petrochemicals, detergents, plastics, paper pollution, etc. Waste waters can be contaminated by feedstock materials, by-products, product material in soluble or particulate form, washing and cleaning agents, solvents and added value products such as plasticisers. Treatment facilities that do not need control of their effluent typically opt for a type of aerobic treatment, i.e. aerated lagoons.

## Electric Power Plants

Fossil-fuel power stations, particularly coal-fired plants, are a major source of industrial wastewater. Many of these plants discharge wastewater with significant levels of metals such as lead, mercury, cadmium and chromium, as well as arsenic, selenium and nitrogen compounds (nitrates and nitrites). Wastewater streams include flue-gas desulfurization, fly ash, bottom ash and flue gas mercury control. Plants with air pollution controls such as wet scrubbers typically transfer the captured pollutants to the wastewater stream.

Ash ponds, a type of surface impoundment, are a widely used treatment technology at coal-fired plants. These ponds use gravity to settle out large particulates (measured as total suspended solids) from power plant wastewater. This technology does not treat dissolved pollutants. Power stations use additional technologies to control pollutants, depending on the particular wastestream in the plant. These include dry ash handling, closed-loop ash recycling, chemical precipitation, biological treatment (such as an activated sludge process), and evaporation.

## Food Industry

Wastewater generated from agricultural and food operations has distinctive characteristics that set it apart from common municipal wastewater managed by public or private sewage treatment plants throughout the world: it is biodegradable and non-toxic, but has high concentrations of biochemical oxygen demand (BOD) and suspended solids (SS). The constituents of food and agriculture wastewater are often complex to predict, due to the differences in BOD and pH in effluents from vegetable, fruit, and meat products and due to the seasonal nature of food processing and post-harvesting.

Processing of food from raw materials requires large volumes of high grade water. Vegetable washing generates waters with high loads of particulate matter and some dissolved organic matter. It may also contain surfactants.

Animal slaughter and processing produces very strong organic waste from body fluids, such as blood, and gut contents. This wastewater is frequently contaminated by significant levels of antibiotics and growth hormones from the animals and by a variety of pesticides used to control external parasites.

Processing food for sale produces wastes generated from cooking which are often rich in plant organic material and may also contain salt, flavourings, colouring material and acids or alkali. Very significant quantities of oil or fats may also be present.

## Iron and Steel Industry

The production of iron from its ores involves powerful reduction reactions in blast furnaces. Cooling waters are inevitably contaminated with products especially ammonia and cyanide. Production of coke from coal in coking plants also requires water cooling and the use of water in by-products separation. Contamination of waste streams includes gasification products such as benzene, naphthalene, anthracene, cyanide, ammonia, phenols, cresols together with a range of more complex organic compounds known collectively as polycyclic aromatic hydrocarbons (PAH).

The conversion of iron or steel into sheet, wire or rods requires hot and cold mechanical transformation stages frequently employing water as a lubricant and coolant. Contaminants include hydraulic oils, tallow and particulate solids. Final treatment of iron and steel products before onward sale into manufacturing includes *pickling* in strong mineral acid to remove rust and prepare the surface for tin or chromium plating or for other surface treatments such as galvanisation or painting. The two acids commonly used are hydrochloric acid and sulfuric acid. Wastewaters include acidic rinse waters together with waste acid. Although many plants operate acid recovery plants (particularly those using hydrochloric acid), where the mineral acid is boiled away from the iron salts, there remains a large volume of highly acid ferrous sulfate or ferrous chloride to be disposed of. Many steel industry wastewaters are contaminated by hydraulic oil, also known as *soluble oil.*

## Mines and Quarries

The principal waste-waters associated with mines and quarries are slurries of rock particles in water. These arise from rainfall washing exposed surfaces and haul roads and also from rock washing and grading processes. Volumes of water can be very high, especially rainfall related arisings on large sites. Some specialized separation operations, such as coal washing to separate coal from native rock using density gradients, can produce wastewater contaminated by fine particulate haematite and surfactants. Oils and hydraulic oils are also common contaminants.

Mine wastewater effluent in Peru, with neutralized pH from tailing runoff.

Wastewater from metal mines and ore recovery plants are inevitably contaminated by the minerals present in the native rock formations. Following crushing and extraction of the desirable materials, undesirable materials may enter the wastewater stream. For metal mines, this can include unwanted metals such as zinc and other materials such as arsenic. Extraction of high value metals such as gold and silver may generate slimes containing very fine particles in where physical removal of contaminants becomes particularly difficult.

Additionally, the geologic formations that harbour economically valuable metals such as copper and gold very often consist of sulphide-type ores. The processing entails grinding the rock into fine particles and then extracting the desired metal(s), with the leftover rock being known as tailings. These tailings contain a combination of not only undesirable leftover metals, but also sulphide components which eventually form sulphuric acid upon the exposure to air and water that inevitably occurs when the tailings are disposed of in large impoundments. The resulting acid mine drainage, which is often rich in heavy metals (because acids dissolve metals), is one of the many environmental impacts of mining.

## Nuclear Industry

The waste production from the nuclear and radio-chemicals industry is dealt with as *Radioactive waste*.

## Pulp and Paper Industry

Effluent from the pulp and paper industry is generally high in suspended solids and BOD. Plants that bleach wood pulp for paper making may generate chloroform, dioxins (including 2,3,7,8-TCDD), furans, phenols and chemical oxygen demand (COD). Stand-alone paper mills using imported pulp may only require simple primary treatment, such as sedimentation or dissolved air flotation. Increased BOD or COD loadings, as well as organic pollutants, may require biological treatment such as activated sludge or upflow anaerobic sludge blanket reactors. For mills with high inorganic loadings like salt, tertiary treatments may be required, either general membrane treatments like ultrafiltration or reverse osmosis or treatments to remove specific contaminants, such as nutrients.

## Industrial Oil Contamination

Industrial applications where oil enters the wastewater stream may include vehicle wash bays, workshops, fuel storage depots, transport hubs and power generation. Often the wastewater is discharged into local sewer or trade waste systems and must meet local environmental specifications. Typical contaminants can include solvents, detergents, grit. lubricants and hydrocarbons.

## Water Treatment

Many industries have a need to treat water to obtain very high quality water for demanding purposes such as environmental discharge compliance. Water treatment produces organic and mineral sludges from filtration and sedimentation. Ion exchange using natural or synthetic resins removes calcium, magnesium and carbonate ions from water, typically replacing them with sodium, chloride, hydroxyl and/or other ions. Regeneration of ion exchange columns with strong acids and alkalis produces a wastewater rich in hardness ions which are readily precipitated out, especially when in admixture with other wastewater constituents.

## Wool Processing

Insecticide residues in fleeces are a particular problem in treating waters generated in wool processing. Animal fats may be present in the wastewater, which if not contaminated, can be recovered for the production of tallow or further rendering.

## Treatment of Industrial Wastewater

The various types of contamination of wastewater require a variety of strategies to remove the contamination.

## Brine Treatment

Brine treatment involves removing dissolved salt ions from the waste stream. Although similarities to seawater or brackish water desalination exist, industrial brine treatment may contain unique combinations of dissolved ions, such as hardness ions or other metals, necessitating specific processes and equipment.

Brine treatment systems are typically optimized to either reduce the volume of the final discharge for more economic disposal (as disposal costs are often based on volume) or maximize the recovery of fresh water or salts. Brine treatment systems may also be optimized to reduce electricity consumption, chemical usage, or physical footprint.

Brine treatment is commonly encountered when treating cooling tower blowdown, produced water from steam assisted gravity drainage (SAGD), produced water from natural gas extraction such as coal seam gas, frac flowback water, acid mine or acid rock drainage, reverse osmosis reject, chlor-alkali wastewater, pulp and paper mill effluent, and waste streams from food and beverage processing.

Brine treatment technologies may include: membrane filtration processes, such as reverse osmosis; ion exchange processes such as electrodialysis or weak acid cation exchange; or evaporation processes, such as brine concentrators and crystallizers employing mechanical vapour recompression and steam.

Reverse osmosis may not be viable for brine treatment, due to the potential for fouling caused by hardness salts or organic contaminants, or damage to the reverse osmosis membranes from hydrocarbons.

Evaporation processes are the most widespread for brine treatment as they enable the highest degree of concentration, as high as solid salt. They also produce the highest purity effluent, even distillate-quality. Evaporation processes are also more tolerant of organics, hydrocarbons, or hardness salts. However, energy consumption is high and corrosion may be an issue as the prime mover is concentrated salt water. As a result, evaporation systems typically employ titanium or duplex stainless steel materials.

## Brine Management

Brine management examines the broader context of brine treatment and may include consideration of government policy and regulations, corporate sustainability, environmental impact, recycling, handling and transport, containment, centralized compared to on-site treatment, avoidance and reduction, technologies, and economics. Brine management shares some issues with leachate management and more general waste management.

## Solids Removal

Most solids can be removed using simple sedimentation techniques with the solids recovered as slurry or sludge. Very fine solids and solids with densities close to the density of water pose special problems. In such case filtration or ultrafiltration may be required. Although, flocculation may be used, using alum salts or the addition of polyelectrolytes.

## Oils and Grease Removal

The effective removal of oils and grease is dependent on the characteristics of the oil in terms of its suspension state and droplet size, which will in turn affect the choice of separator technology.

Oil pollution in water usually comes in four states, often in combination:

- free oil - large oil droplets sitting on the surface;
- heavy oil, which sits at the bottom, often adhering to solids like dirt;
- emulsified, where the oil droplets are heavily "chopped"; and
- dissolved oil, where the droplets are fully dispersed and not visible. Emulsified oil droplets are the most common in industrial oily wastewater and are extremely difficult to separate.

The methodology for separating the oil is dependent on the oil droplet size. Larger oil droplets such as those in free oil pollution are easily removed, but as the droplets become smaller, some separator technologies perform better than others.

Most separator technologies will have an optimum range of oil droplet sizes that can be effectively treated. This is known as the "micron rating."

Analysing the oily water to determine droplet size can be performed with a video particle analyser. Alternatively, there are commonalities in industries for oil droplet sizes.

Larger droplets–greater than 60 microns–are often present in wastewater in workshops, re-fuel areas and depots. Twenty to 50 micron oil droplets often are present in vehicle wash bays, meat processing and dairy manufacturing effluent and aluminium billet cooling towers. Smaller droplets in the range of 10 to 20 microns tend to occur in workshops and condensates.

Each separator technology will have its' own performance curve outlining optimum performance based on oil droplet size. the most common separators are gravity tanks or pits, API oil-water separators or plate packs, chemical treatment via DAFs, centrifuges, media filters and hydrocyclones.

## APi Separators

Many oils can be recovered from open water surfaces by skimming devices. Considered a dependable and cheap way to remove oil, grease and other hydrocarbons from water, oil skimmers can sometimes achieve the desired level of water purity. At other times, skimming is also a cost-efficient method to remove most of the oil before using membrane filters and chemical processes. Skimmers will prevent filters from blinding prematurely and keep chemical costs down because there is less oil to process.

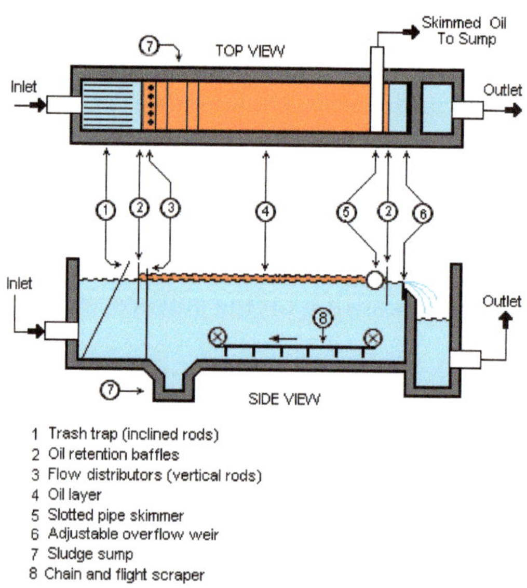


A typical API oil-water separator used in many industries

Because grease skimming involves higher viscosity hydrocarbons, skimmers must be equipped with heaters powerful enough to keep grease fluid for discharge. If floating grease forms into solid clumps or mats, a spray bar, aerator or mechanical apparatus can be used to facilitate removal.

However, hydraulic oils and the majority of oils that have degraded to any extent will also have a soluble or emulsified component that will require further treatment to eliminate. Dissolving or emulsifying oil using surfactants or solvents usually exacerbates the problem rather than solving it, producing wastewater that is more difficult to treat.

The wastewaters from large-scale industries such as oil refineries, petrochemical plants, chemical plants, and natural gas processing plants commonly contain gross amounts of oil and suspended solids. Those industries use a device known as an API oil-water separator which is designed to separate the oil and suspended solids from their wastewater effluents. The name is derived from the fact that such separators are designed according to standards published by the American Petroleum Institute (API).

The API separator is a gravity separation device designed by using Stokes Law to define the rise velocity of oil droplets based on their density and size. The design is based on the specific gravity difference between the oil and the wastewater because that difference is much smaller than the specific gravity difference between the suspended solids and water. The suspended solids settles to the bottom of the separator as a sediment layer, the oil rises to top of the separator and the cleansed wastewater is the middle layer between the oil layer and the solids.

Typically, the oil layer is skimmed off and subsequently re-processed or disposed of, and the bottom sediment layer is removed by a chain and flight scraper (or similar device) and a sludge pump. The water layer is sent to further treatment for additional removal of any residual oil and then to some type of biological treatment unit for removal of undesirable dissolved chemical compounds.

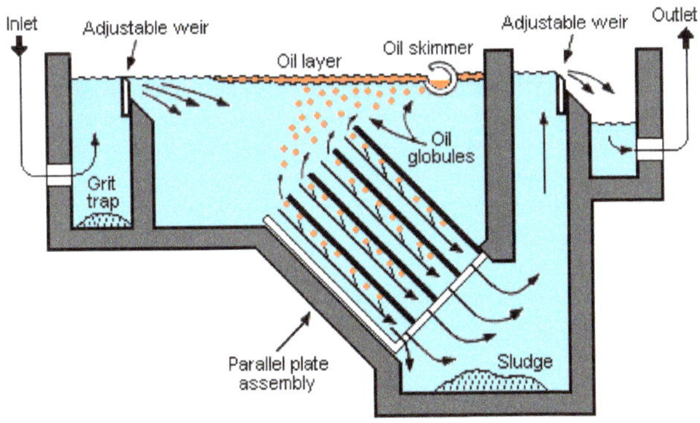


A typical parallel plate separator

Parallel plate separators are similar to API separators but they include tilted parallel plate assemblies (also known as parallel packs). The parallel plates provide more surface for suspended oil droplets to coalesce into larger globules. Such separators still depend upon the specific gravity between the suspended oil and the water. However, the parallel plates enhance the degree of oil-water separation. The result is that a parallel plate separator requires significantly less space than a conventional API separator to achieve the same degree of separation.

Hydrocyclone oil separators

Hydrocyclone oil separators operate on the process where wastewater enters the cyclone chamber and is spun under extreme centrifugal forces more than 1000 times the force of gravity. This force causes the water and oil droplets to separate. The separated oil is discharged from one end of the cyclone where treated water is discharged through the opposite end for further treatment, filtration or discharge.

Hydrocyclones are useful for the greatest range of oil droplet sizes operating from less than 10 microns and up and can operate continuously without water pre-treatment and at any temperature and pH. Applications where hydrocyclones are found are in industry where oily water sources arise in workshops, vehicle wash bays, transport hubs, fuel depots and aluminium billet processing. Animal fats from meat processing and dairy manufacturing can also be removed without the need of chemical treatment that often is required for dissolved air flotation (DAF) systems.

## Removal of Biodegradable Organics

Biodegradable organic material of plant or animal origin is usually possible to treat using extended conventional sewage treatment processes such as activated sludge or trickling filter. Problems can arise if the wastewater is excessively diluted with washing water or is highly concentrated such as undiluted blood or milk. The presence of cleaning agents, disinfectants, pesticides, or antibiotics can have detrimental impacts on treatment processes.

## Activated Sludge Process

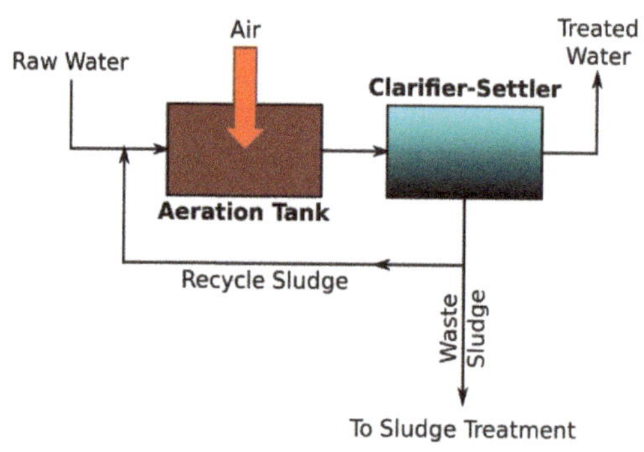


A generalized diagram of an activated sludge process.

Activated sludge is a biochemical process for treating sewage and industrial wastewater that uses air (or oxygen) and microorganisms to biologically oxidize organic pollutants, producing a waste sludge (or floc) containing the oxidized material. In general, an activated sludge process includes:

- An aeration tank where air (or oxygen) is injected and thoroughly mixed into the wastewater.

- A settling tank (usually referred to as a clarifier or "settler") to allow the waste sludge to settle. Part of the waste sludge is recycled to the aeration tank and the remaining waste sludge is removed for further treatment and ultimate disposal.

## Trickling Filter Process

A trickling filter consists of a bed of rocks, gravel, slag, peat moss, or plastic media over which wastewater flows downward and contacts a layer (or film) of microbial slime covering the bed media. Aerobic conditions are maintained by forced air flowing through the bed or by natural convection of air. The process involves adsorption of organic compounds in the wastewater by the microbial slime layer, diffusion of air into the slime layer to provide the oxygen required for the biochemical oxidation of the organic compounds. The end products include carbon dioxide gas, water and other products of the oxidation. As the slime layer thickens, it becomes difficult for the air to penetrate the layer and an inner anaerobic layer is formed.

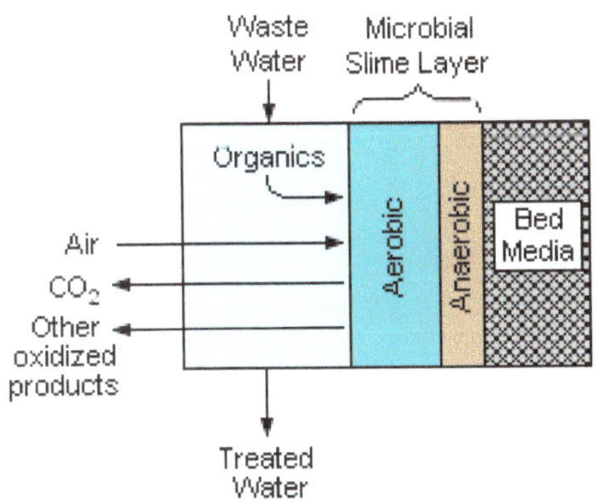


A schematic cross-section of the contact face of the bed media in a trickling filter

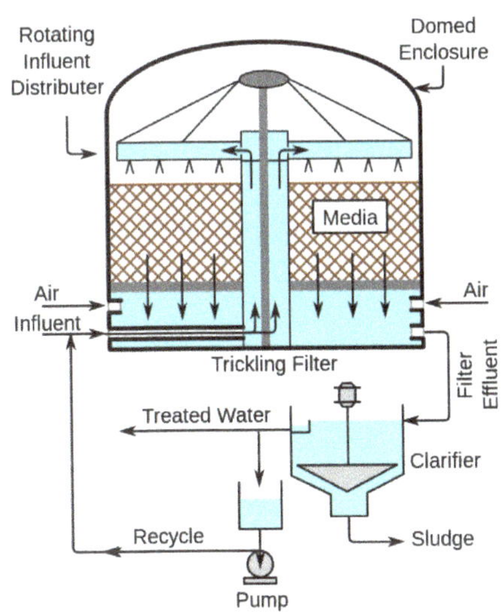


A typical complete trickling filter system

The fundamental components of a complete trickling filter system are:

- A bed of filter medium upon which a layer of microbial slime is promoted and developed.
- An enclosure or a container which houses the bed of filter medium.
- A system for distributing the flow of wastewater over the filter medium.
- A system for removing and disposing of any sludge from the treated effluent.

The treatment of sewage or other wastewater with trickling filters is among the oldest and most well characterized treatment technologies.

A trickling filter is also often called a *trickle filter, trickling biofilter, biofilter, biological filter* or *biological trickling filter*.

## Treatment of Other Organics

Synthetic organic materials including solvents, paints, pharmaceuticals, pesticides, products from coke production and so forth can be very difficult to treat. Treatment methods are often specific to the material being treated. Methods include advanced oxidation processing, distillation, adsorption, vitrification, incineration, chemical immobilisation or landfill disposal. Some materials such as some detergents may be capable of biological degradation and in such cases, a modified form of wastewater treatment can be used.

## Treatment of Acids and Alkalis

Acids and alkalis can usually be neutralised under controlled conditions. Neutralisation frequently produces a precipitate that will require treatment as a solid residue that may also be toxic. In some cases, gases may be evolved requiring treatment for the gas stream. Some other forms of treatment are usually required following neutralisation.

Waste streams rich in hardness ions as from de-ionisation processes can readily lose the hardness ions in a buildup of precipitated calcium and magnesium salts. This precipitation process can cause severe *furring* of pipes and can, in extreme cases, cause the blockage of disposal pipes. A 1 metre diameter industrial marine discharge pipe serving a major chemicals complex was blocked by such salts in the 1970s. Treatment is by concentration of de-ionisation waste waters and disposal to landfill or by careful pH management of the released wastewater.

## Treatment of Toxic Materials

Toxic materials including many organic materials, metals (such as zinc, silver, cadmium, thallium, etc.) acids, alkalis, non-metallic elements (such as arsenic or selenium) are generally resistant to biological processes unless very dilute. Metals can often be precipitated out by changing the pH or by treatment with other chemicals. Many, however, are resistant to treatment or mitigation and may require concentration followed by landfilling or recycling. Dissolved organics can be *incinerated* within the wastewater by the advanced oxidation process.

# Vehicle Emissions Control

Vehicle emissions control is the study of reducing the motor vehicle emissions—emissions produced by motor vehicles, especially internal combustion engines.

## Types of Emissions

Emissions of many air pollutants have been shown to have variety of negative effects on public health and the natural environment. Emissions that are principal pollutants of concern include:

- Hydrocarbons - A class of burned or partially burned fuel, hydrocarbons are toxins. Hydrocarbons are a major contributor to smog, which can be a major problem in urban areas. Prolonged exposure to hydrocarbons contributes to asthma, liver disease, lung disease, and cancer. Regulations governing hydrocarbons vary according to type of engine and jurisdiction; in some cases, "non-methane hydrocarbons" are regulated, while in other cases, "total hydrocarbons" are regulated. Technology for one application (to meet a non-methane hydrocarbon standard) may not be suitable for use in an application that has to meet a total hydrocarbon standard. Methane is not directly toxic, but is more difficult to break down in a catalytic converter, so in effect a "non-methane hydrocarbon" regulation can be considered easier to meet. Since methane is a greenhouse gas, interest is rising in how to eliminate emissions of it.

- Carbon monoxide (CO) - A product of incomplete combustion, carbon monoxide reduces the blood's ability to carry oxygen; overexposure (carbon monoxide poisoning) may be fatal. Carbon Monoxide poisoning is a killer in high concentrations.

- $NO_x$ - Generated when nitrogen in the air reacts with oxygen at the high temperature and pressure inside the engine. $NO_x$ is a precursor to smog and acid rain. $NO_x$ is the sum of NO and $NO_2$. $NO_2$ is extremely reactive. $NO_x$ production is increased when an engine runs at its most efficient (i.e. hottest) part of the cycle.

- Particulate matter – Soot or smoke made up of particles in the micrometre size range: Particulate matter causes negative health effects, including but not limited to respiratory disease and cancer.

- Sulfur oxide ($SO_x$) - A general term for oxides of sulfur, which are emitted from motor vehicles burning fuel containing sulfur. Reducing the level of fuel sulfur reduces the level of Sulfur oxide emitted from the tailpipe.
- Volatile organic compounds (VOCs) - Organic compounds which typically have a boiling point less than or equal to 250 °C; for example chlorofluorocarbons (CFCs) and formaldehyde. Volatile organic compounds are a subsection of Hydrocarbons that are mentioned separately because of their dangers to public health.

## History

Throughout the 1950s and 1960s, various federal, state and local governments in the United States conducted studies into the numerous sources of air pollution. These studies ultimately attributed a significant portion of air pollution to the automobile, and concluded air pollution is not bounded by local political boundaries. At that time, such minimal emission control regulations as existed in the U.S. were promulgated at the municipal or, occasionally, the state level. The ineffective local regulations were gradually supplanted by more comprehensive state and federal regulations. By 1967 the State of California created the California Air Resources Board, and in 1970, the federal United States Environmental Protection Agency (EPA) was established. Both agencies, as well as other state agencies, now create and enforce emission regulations for automobiles in the United States. Similar agencies and regulations were contemporaneously developed and implemented in Canada, Western Europe, Australia, and Japan.

The first effort at controlling pollution from automobiles was the PCV (positive crankcase ventilation) system. This draws crankcase fumes heavy in unburned hydrocarbons — a precursor to photochemical smog — into the engine's intake tract so they are burned rather than released unburned from the crankcase into the atmosphere. Positive crankcase ventilation was first installed on a widespread basis by law on all new 1961-model cars first sold in California. The following year, New York required it. By 1964, most new cars sold in the U.S. were so equipped, and PCV quickly became standard equipment on all vehicles worldwide.

The first legislated exhaust (tailpipe) emission standards were promulgated by the State of California for 1966 model year for cars sold in that state, followed by the United States as a whole in model year 1968. Also in 1966, the first emission test cycle was enacted in the State of California measuring tailpipe emissions in PPM (parts per million). The standards were progressively tightened year by year, as mandated by the EPA.

By the 1974 model year, the emission standards had tightened such that the de-tuning techniques used to meet them were seriously reducing engine efficiency and thus increasing fuel usage. The new emission standards for 1975 model year, as well as the increase in fuel usage, forced the invention of the catalytic converter for after-treatment of the exhaust gas. This was not possible with existing leaded gasoline, because the lead residue contaminated the platinum catalyst. In 1972, General Motors proposed to the American Petroleum Institute the elimination of leaded fuels for 1975 and later model year cars. The production and distribution of unleaded fuel was a major challenge, but it was completed successfully in time for the 1975 model year cars. All modern cars are now equipped with catalytic converters and leaded fuel is nearly impossible to buy in most First World countries.

## Regulatory Agencies

The agencies charged with regulating exhaust emissions vary from jurisdiction to jurisdiction, even in the same country. For example, in the United States, overall responsibility belongs to the EPA, but due to special requirements of the State of California, emissions in California are regulated by the Air Resources Board. In Texas, the Texas Railroad Commission is responsible for regulating emissions from LPG-fueled rich burn engines (but not gasoline-fueled rich burn engines).

## North America

- California Air Resources Board - California, United States (most sources)
- Environment Canada - Canada (most sources)
- Environmental Protection Agency - United States (most sources)
- Texas Railroad Commission - Texas, United States (LPG-fueled engines only)
- Transport Canada - Canada (trains and ships)

## Europe

The European Union has control over regulation of emissions in EU member states; however, many member states have their own government bodies to enforce and implement these regulations in their respective countries. In short, the EU forms the policy (by setting limits such as the European emission standard) and the member states decide how to best implement it in their own country.

## United Kingdom

In the United Kingdom, matters concerning environmental policy are what is known as "devolved powers" which means, each of the constituent countries deals with it separately through their own government bodies set up to deal with environmental issues in their respective country:

- Environment Agency - England and Wales

- Scottish Environment Protection Agency (SEPA) - Scotland
- Department of the Environment - Northern Ireland

However, many UK-wide policies are handled by the Department of the Environment Food and Rural Affairs (DEFRA) and they are still subject to EU regulations.

## Emissions Control

Engine efficiency has been steadily improved with improved engine design, more precise ignition timing and electronic ignition, more precise fuel metering, and computerized engine management.

Advances in engine and vehicle technology continually reduce the toxicity of exhaust leaving the engine, but these alone have generally been proved insufficient to meet emissions goals. Therefore, technologies to detoxify the exhaust are an essential part of emissions control.

## Air Injection

One of the first-developed exhaust emission control systems is secondary air injection. Originally, this system was used to inject air into the engine's exhaust ports to provide oxygen so unburned and partially burned hydrocarbons in the exhaust would finish burning. Air injection is now used to support the catalytic converter's oxidation reaction, and to reduce emissions when an engine is started from cold. After a cold start, an engine needs an air-fuel mixture richer than what it needs at operating temperature, and the catalytic converter does not function efficiently until it has reached its own operating temperature. The air injected upstream of the converter supports combustion in the exhaust headpipe, which speeds catalyst warmup and reduces the amount of unburned hydrocarbon emitted from the tailpipe.

## Exhaust Gas Recirculation

In the United States and Canada, many engines in 1973 and newer vehicles (1972 and newer in California) have a system that routes a metered amount of exhaust into the intake tract under particular operating conditions. Exhaust neither burns nor supports combustion, so it dilutes the air/fuel charge to reduce peak combustion chamber temperatures. This, in turn, reduces the formation of $NO_x$.

## Catalytic Converter

The catalytic converter is a device placed in the exhaust pipe, which converts hydrocarbons, carbon monoxide, and $NO_x$ into less harmful gases by using a combination of platinum, palladium and rhodium as catalysts.

There are two types of catalytic converter, a two-way and a three-way converter. Two-way converters were common until the 1980s, when three-way converters replaced them on most automobile engines.

### Evaporative Emissions Control

Evaporative emissions are the result of gasoline vapors escaping from the vehicle's fuel system. Since 1971, all U.S. vehicles have had fully sealed fuel systems that do not vent directly to the atmosphere; mandates for systems of this type appeared contemporaneously in other jurisdictions. In a typical system, vapors from the fuel tank and carburetor bowl vent (on carbureted vehicles) are ducted to canisters containing activated carbon. The vapors are adsorbed within the canister, and during certain engine operational modes fresh air is drawn through the canister, pulling the vapor into the engine, where it burns.

### Remote Sensing Emission Testing - Field Studies

Some US states are also using a technology developed by Dr. Donald H. Stedman of the University of Denver, which uses infra-red and ultraviolet light to detect emissions while vehicles pass by on public roads, thus eliminating the need for owners to go to a test center. Stedman's invisible light flash detection of exhaust gases is commonly used in metropolitan areas, is offered by the US-Swedish company OPUS Inspection and becoming more broadly known in Europe.

### Use of Emission Test Data

Emission test results from individual vehicles are in many cases compiled to evaluate the emissions performance of various classes of vehicles, the efficacy of the testing program and of various other emission-related regulations (such as changes to fuel formulations) and to model the effects of auto emissions on public health and the environment. For example, the Environmental Working Group used California ASM emissions data to create an "Auto Asthma Index" that rates vehicle models according to emissions of hydrocarbons and nitrogen oxides, chemical precursors to photochemical smog.

## Ecological Sanitation

Ecological sanitation, commonly abbreviated to ecosan (also spelled eco-san or EcoSan), is an approach which is characterized by a desire to safely "close the loop" (mainly for the nutrients and organic matter) between sanitation and agriculture. Ecosan systems safely recycle excreta resources (plant nutrients and organic matter) to crop production in such a way that the use of non-renewable resources is minimised. When properly designed and operated, ecosan systems can strive to provide a hygienically safe, economical, and closed-loop system to convert human excreta into nutrients to be returned to the soil, and water to be returned to the land.

Ecosan concept showing a separation of flow streams, treatment and reuse

## Definition

The definitions of ecosan have varied in the past. Nowadays, the most widely accepted definition of ecosan, formulated by experts in Sweden, is: "Ecological sanitation systems are systems which allow for the safe recycling of nutrients to crop production in such a way that the use of non-renewable resources is minimized. These systems have a strong potential to be sustainable sanitation systems if technical, institutional, social and economic aspects are managed appropriately."

Traditionally, ecosan has often been associated with urine diversion and in particular with urine-diverting dry toilets (UDDTs), a type of dry toilet. For this reason, the term "ecosan toilet" is widely used when people mean a UDDT. However, the ecosan concept should not be limited to one particular type of toilet. Also, UDDTs can be used without having any re-use activities in which case they are not in line with the ecosan concept (an example being the 80,000 UDDTs implemented by eThekwini Municipality near Durban, South Africa).

## Overview

Poster by EcoSanRes program: Closing the loop on Sanitation (2005)

Ecosan closing the loop poster (in French), by the NGO CREPA in 2005, UDDTs are used in this example

## Objectives

The main objectives of ecological sanitation are to reduce the health risks related to sanitation, contaminated water and waste; to prevent groundwater pollution and surface water pollution; and to reuse nutrients or energy contained within wastes.

## Resource Recovery

The statement in the definition of ecosan to "safely recycle" includes hygienic, microbial and chemical aspects. Thus, the recycled human excreta product, in solid or liquid form, shall be of high quality both concerning pathogens and all kind of hazardous chemical components. The statement "use of non-renewable resources is minimised" means that the gain in resources by recycling shall be larger than the cost of resources by recycling.

Ecosan is based on an overall concept of material flows as part of an ecologically and economically sustainable wastewater management system tailored to the needs of the users and to the respective local conditions. It does not favour a specific sanitation technology, but is rather a certain philosophy in handling substances that have so far been seen simply as wastewater and water-carried waste for disposal.

## Reuse as Fertilizer

Right from the start, the first proponents of ecosan systems had a strong focus on increasing agricultural productivity (via the reuse of excreta as fertilizers) and thus improving the nutritional status of the people at the same time as providing them with safe sanitation. Disease reduction was meant to be achieved not only by reducing infections transmitted via the fecal-oral route but also by reducing malnutrition in children. This link between WASH, nutrition, a disease called environmental enteropathy (or tropical enteropathy) as well as stunted growth of children has risen to the top of the agenda of the WASH sector since about 2013.

Agricultural trials around the world have shown measurable benefits of using treated excreta in agriculture as a fertilizer and soil conditioner. This applies in particular to the use of urine. Reuse trials in Zimbabwe showed positive results for using urine on green, leafy plants such as spinach or maize as well as fruit trees. Another study in Finland indicated that the use of urine and the use of urine and wood ash "could produce 27% and 10% more red beet root biomass." Urine has been proven in many studies to be a valuable, relatively easy to handle fertilizer, containing nitrogen, phosphorus, potassium and important micro-nutrients.

## Phosphorus Recovery

Another aspect that ecosan systems are trying to address is the possible upcoming shortage of phosphorus. Phosphorus has an important role for plant growth, and therefore in fertilizer production, but is a limited mineral resource. The situation is similar for potassium. Known mineral phosphate rock reserves are becoming scarce and increasingly costly to extract - this is also called the "peak phosphorus" crisis. One review of global phosphate supply suggested that if collected, phosphate in urine could supply 22% of the total demand.

## Comparison With Sustainable Sanitation

The definition of ecosan is focusing on the health, environment and resource aspect of sustainable sanitation. Thus ecosan is not, per se, sustainable sanitation, but ecosan systems can be implemented in a sustainable way and have a strong potential for sustainable sanitation, if technical, institutional, social and economical aspects are cared for appropriately. Ecosan systems can be "unsustainable" for example if there is too little user acceptance or if the costs of the system are too high for a given target group of users, making the system financially unsustainable in the longer term.

## Possible Benefits

Benefits of ecosan systems may include:

- Minimizing the introduction of pathogens from human excreta into the water cycle (groundwater and surface water) - a major consideration in low-lying geographies is groundwater pollution by pit latrines. In many areas where the water table is high, pit latrines directly pollute the water table, potentially affecting the large numbers of people.
- Promotion of safe, hygienic recovery and use of nutrients (nitrogen and phosphorus), organics, trace elements, water and energy
- Promotion of soil fertility, improvement of agricultural productivity and food security
- Contribution to the conservation of resources through lower water consumption, substitution of mineral fertilizer and minimization of water pollution
- Less reliance on mined phosphorus for fertilizer production

- Energy reduction in fertilizer production: Urea is the major component of urine, yet we produce vast quantities of urea by using fossil fuels. By properly managing urine, treatment costs as well as fertilizer costs can be reduced.

## Challenges

The ecosan approach has been criticised for being overly focused on reuse in agriculture, whilst neglecting some of the other criteria for sustainable sanitation. In fact, ecosan systems can be "unsustainable" for example if there is too little user acceptance or if the costs of the system are too high for a given target group of users, making the system financially unsustainable in the longer term.

Some proponents of ecosan have been criticised as being too dogmatic, with an over-emphasis on environmental resource protection rather than a focus on public health protection and provision of sanitation at a very low cost (for example UDDTs, which some people call "ecosan toilets" may be more expensive to build than pit latrines, even if in the longer term they are cheaper to maintain).

The safety of ecosan systems in terms of pathogen destruction during the various treatment processes is a continuous topic of debate between proponents and opponents of ecosan systems. However, the publication of the WHO Guidelines on Reuse, with its multiple barrier concept, has gone a long way in establishing a common framework for safe reuse. Nevertheless, the question remains whether ecosan systems can ever be scaled up to reach millions of people and how they can be made sufficiently safe to operate. The initial excitement in the early 2000s by the ecosan pioneers has changed into a realization that changing attitudes and behaviours in sanitation takes a lot of patience.

Acknowledgement for ecosan came with the awarding of the Stockholm Water Prize in 2013 to Peter Morgan, a pioneer of handpumps and ventilated pit latrines (VIPs) in addition to ecosan-type toilets (the Arborloo, the Skyloo and the Fossa alterna). Peter Morgan is renowned as one of the leading creators and proponents of ecological sanitation solutions, which enable the safe reuse of human excreta to enhance soil quality and crop production. His ecosan-type toilets are now in use in countries across the globe, centred on converting a sanitary problem into a productive resource.

Also many of the research projects that the Bill and Melinda Gates Foundation have been funding since about 2011 in sanitation are dealing with resource recovery - this might well be a legacy of the ecosan concept, even if the term "ecosan" is not used by these researchers.

## Technologies Used in Ecosan Systems

Ecosan offers a flexible framework, where centralised elements can be combined with decentralised ones, waterborne with dry sanitation, high-tech with low-tech, etc. By considering a much larger range of options, optimal and economic solutions can be developed for each particular situation. Technologies used in ecosan systems often - but not always - include elements of source separation, i.e. keeping different waste streams separate, as this can make treatment and safe reuse easier.

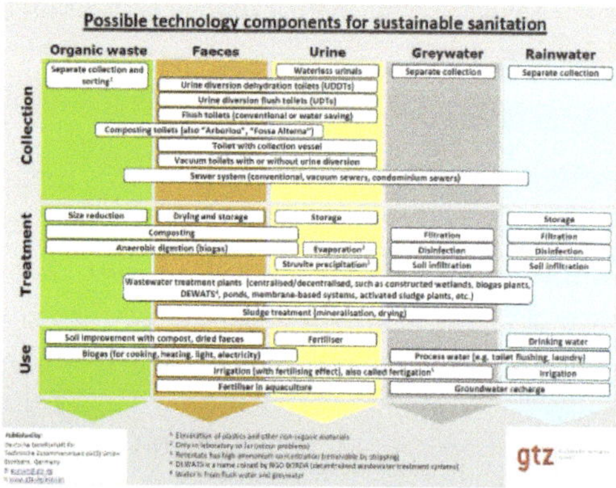

Possible technology components for sustainable sanitation, of which ecosan is a sub-set focussing on the reuse possibilities

The most common technology used in ecosan systems is the urine-diverting dry toilet, but ecosan systems can also use other technologies, such as vacuum toilets coupled with biogas plants, constructed wetlands, composting toilets and so forth.

Examples of ecosan projects worldwide can be found in a list published by GIZ in 2012, as well as in those casc studies published by the Sustainable Sanitation Alliance that are focused on reuse activities.

## History

### Excreta Reuse in Dry Sanitation Systems

The recovery and use of urine and feces in "dry sanitation systems", i.e. without sewers or without mixing substantial amounts of water with the excreta, has been practiced by almost all cultures. The reuse was not limited to agricultural production. The Romans, for example, were aware of the bleaching attribute of the ammonia within urine and used it to whiten clothing.

Many traditional agricultural societies recognized the value of human waste for soil fertility and practised the "dry" collection and reuse of excreta. This enabled them to live in communities in which nutrients and organic matter contained in excreta were returned to the soil. Historical descriptions about these practices are sparse, but it is known that excreta reuse was practiced widely in Asia (for example in China, Japan, Vietnam, Cambodia, Korea) but also in Central and South America. However, the most renowned example of the organised collection and use of human excreta to support food production is that of China. The value of "night soil" as a fertilizer was recognized with well-developed systems in place to enable the collection of excreta from cities and its transportation to fields. The Chinese were aware of the benefits of using excreta in crop production more than 2500 years ago, enabling them to sustain more people at a higher density than any other system of agriculture.

In Mexico the Aztec culture collected human excreta for agricultural use. One example of this practice has been documented for the Aztec city of Tenochtitlan which was founded in 1325 and was one of the last cities of pre-Hispanic Mexico (conquered in 1521 by the Spanish): The population placed the sweepings in special boats moored at docks around the city. Mixtures of sweepings and excreta were used to fertilize the chinampas (agricultural fields) or to bolster the banks bordering the lake. Urine was collected in containers in all houses, then mixed with mud and used as a fabric dye. The Aztecs recognized the importance of recycling nutrients and compounds contained in wastewater.

In Peru, the Incas had a high regard for excreta as a fertilizer, which was stored, dried and pulverized to be utilized when planting maize.

In the Middle Ages, the use of excreta and greywater in agricultural production was the norm. European cities were rapidly urbanizing and sanitation was becoming an increasingly serious problem, whilst at the same time the cities themselves were becoming an increasingly important source of agricultural nutrients. The practice of directly using the nutrients in excreta and wastewater for agriculture therefore continued in Europe into the middle of the 19th Century. Farmers, recognizing the value of excreta, were eager to get these fertilizers to increase production and urban sanitation benefited. This practice was also called gong farmer in England but carried many health risks for those involved with transporting the excreta and fecal sludge.

Traditional forms of sanitation and excreta reuse have continued in various parts of the world for centuries and were still common practice at the advent of the Industrial Revolution. Even as the world became increasingly more urbanised, the nutrients in excreta collected from urban sanitation systems without mixing with water were still used in many societies as a resource to maintain soil fertility, despite rising population densities.

## Decline in Recovery of Nutrients from Human Excreta in Dry Systems

Recovery of nutrients and organic matter from excreta and greywater in non-sewered sanitation systems was addressing the sanitation problems in settlements in Europe and elsewhere and was contributing to securing agricultural productivity. However, the practice did not become the dominant approach to urban sanitation in the 20th century and was gradually replaced with sewer-based sanitation systems without nutrient recovery (apart from agricultural reuse of sewage sludge in some cases) - at least for cities that can afford it. There were four main driving factors that led to the demise in the recovery and use of excreta and greywater from European cities in the 19th century:

- Growth of urban settlements and increasing distance from agricultural fields: urban settlements had grown dramatically over the centuries. The logistical challenge of removing the feces of a booming population from densely packed city centres to bring to agricultural areas many miles away proved too great.

- Increasing water consumption and use of flush toilet: Water flushing greatly increased the volume of sewage, at the same time diluting the nutrients, making it virtually impossible for them to be recovered and reused as they previously were.
- Production of cheap synthetic fertilizers: The nutrient demand of farmers was eventually met by cheap chemical fertilizers, making any efforts to recover and reuse the nutrients and organic material from the large volumes of sewage completely obsolete.
- Political intervention as a consequence of the perceived need for a change with regards to how to deal with odorous substances: Up until the end of the nineteenth century the dominant theory on the spread of illness was the miasma theory. This theory stated that everything that smelled had to be gotten rid off because inhaling bad smells was thought to lead to illness. The miasma theory did contribute to containing disease in urban settlements, but did not allow for a suitable approach to safe excreta reuse to be adopted.

Interestingly, the use of (odorous) animal manure in agriculture continued through to this day, probably because the odour of manure was not thought to contribute to human illnesses and because the vast amount of manure could not be "flushed away" like human excreta can.

Even in situations where dry sanitation systems have been replaced by sewer-based sanitation systems, the recovery of nutrients from wastewater may continue in two forms:

- Wastewater reuse or resource recovery: Use of raw, treated or partially treated wastewater for irrigation in agriculture (with the associated health risks if it is done in an improper way which is often the case in developing countries); and
- Application of sewage sludge to agricultural lands which is not without controversy in many industrialised countries due to the risks of polluting soils with heavy metals and micropollutants if not managed properly.

## Research from 1990s Onwards

The Swedish International Development Cooperation Agency (Sida) funded the "SanRes R&D programme" during 1993 to 2001 which lay the foundation for the subsequent "EcoSanRes programme" carried out by Stockholm Environment Institute (2002–2011). A publication by Sida called "Ecological sanitation" in 1998 compiled the knowledge generated to date about ecosan in a popular book which was published as a second edition in 2004. The book has also been translated into Chinese, French and Spanish.

The German government enterprise GIZ also had a large "ecosan program" from 2001 to 2012. Whilst the term "ecosan" was preferred in the initial stages of this program, it was from 2007 onwards more and more replaced by the broader term "sustainable sanitation". In fact, the Sustainable Sanitation Alliance was founded in 2007 in an attempt to broaden the ecosan concept and to bring together various actors under one umbrella.

Research into how to make reuse of urine and feces safe in agriculture was carried out by Swedish researchers, for example Hakan Jönsson and his team, whose publication on "Guidelines on the Use of Urine and feces in Crop Production" was a milestone which was later incorporated into the WHO "Guidelines on Safe Reuse of Wastewater, Excreta and Greywater" from 2006. The multiple barrier concept to reuse, which is the key cornerstone of this publication, has led to a clear understanding on how excreta reuse can be done safely.

## Workshops and Conferences

Initially, there were dedicated "ecosan conferences" to present and discuss research on ecosan projects:

- A first workshop on ecological sanitation was held in Balingsholm, Sweden in 1997, where all the then established ecosan experts, such as Håkan Jönsson, Peter Morgan (winner of the 2013 Stockholm Water Prize), Ron Sawyer, George Anna Clark and Gunder Edström participated.
- Workshop in Mexico in 1999 with the title "Closing the Loop - Ecological sanitation for food security"
- Ecosan conference in Bonn, Germany in 2000
- First international ecosan conference in Nanning, China in 2001
- Second ecosan conference in Lübeck Germany in 2003
- Third ecosan conference in Durban, South Africa in 2005
- Ecosan conference in Fortaleza, Brazil called "International Conference on Sustainable Sanitation - Water and Food Security for Latin America" in 2007

Since then the ecosan theme has been integrated into other WASH conferences, and separate large ecosan conferences have no longer been organised.

## Use of the Term "Ecosan"

The term "ecosan" was first used in about the 1990s (or perhaps even late 1980s) by an NGO in Ethiopia called Sudea. They used it for urine-diverting dry toilets coupled with reuse activities.

In the ecosan concept, human excreta and wastewater is regarded as a potential resource - which is why it has also been called "resource oriented sanitation". The term "productive sanitation" has also been in use since about 2006.

## Disputes Amongst Experts

In the early days during the 1990s when the term ecosan was something new, discussions were heated and confrontational. Supporters of ecosan claimed the corner on containment, treatment and reuse. The "establishment" defended deep pit latrines, and waterborne systems and they felt offended. Ecosan supporters criticised conventional sanitation for killing children, contaminating waterways with nutrients and pathogens and making helminth worms a global pandemic with upwards of 2 billion cases. Since then, the two opposing sides have slowly found ways of dealing with each other, and the formation of the Sustainable Sanitation Alliance has further helped to provide a space for all sanitation actors to meet and push into the same direction of sustainable sanitation.

## Examples

- Sweden is the leader in Europe to put ecosan into practice at a larger scale. For example, Tanum Municipality in Sweden has introduced urine separation toilets due to their very rocky and challenging terrain initially, and later also to recover phosphorus.
- Sweden has also made it possible in 2013 to certify safe and sanitized blackwater (urine and human excreta) from blackwater systems and for further use as a recognized fertilizer. Such blackwater systems could be vacuum toilets or septic tanks. The criteria for the certification have been developed by the Swedish Institute for Agricultural and Environmental Engineering and may pave the way for farmers to use human waste for agricultural production. The Federation of Swedish Farmers have been active in this development. Furthermore, the Swedish EPA in their last proposal in 2014 has downgraded the hygiene risk associated with urine. Previously the normal storage requirement for hygienic quality for large scale use of urine was 6 months. Now they propose decreasing this to one month.
- Stockholm Environment Institute (SEI) ran a large worldwide ecosan research programme called "Ecosanres" from 2001 to 2011. One of the "dry ecosan" pilot projects (i.e. with using dry toilets) of this programme was a large scale implementation of urine-diverting dry toilets (UDDTs) in multi-story buildings together with other technologies to allow resource recovery from excreta. This project was called the Erdos Eco-Town Project in a town called Erdos in the Inner Mongolia Autonomous Region of China. It was a collaboration between the Dongsheng District government in Erdos and the Stockholm Environ-

ment Institute and aimed to save water and provide sanitation services in this drought-stricken and rapidly urbanizing area of northern China. For a variety of technical, social and institutional reasons, the UDDTs were removed after only a few years and the project failed to deliver in the area of nutrient recovery. This project is now well documented and has raised more awareness of the challenges and disadvantages of "urban ecosan".

- SOIL in Haiti built what they call "ecosan toilets" (UDDTs) as part of the emergency relief effort following the 2010 Haiti earthquake. More than 20,000 Haitians are currently using SOIL ecological sanitation toilets and SOIL has produced over 400,000 liters of compost as a result. The compost is used for agricultural and reforestation projects. SOIL's composting process is effective in inactivating *Ascaris* eggs - an indicator for helminth eggs in general - in the excreta collected from the dry toilets within 16 weeks. The composting and monitoring methos used by SOIL in Haiti may serve as an example for other international settings.
- Wherever the Need, an NGO in the UK build ecosan facilities (UDDTs) in various parts of the developing world. They predominantly work in Tamil Nadu (India), where the Tamil Nadu State Government provides subsidies for their work. Wherever the Need have also constructed ecosan in other parts of rural India, Kenya and Sierra Leone. According to their website, heir ecosan projects have positively affected 50,000 people in the developing world.
- The NGO CREPA which was operating in the French-speaking West Africa region (now bi-lingual and called WSA - Water and Sanitation in Africa) was very active in ecosan promotion during the years of 2002 - 2010 with a strong focus on UDDTs coupled with reuse in agriculture, especially in Burkina Faso.

## Ecological Engineering

Ecological engineering is an emerging study of integrating ecology and engineering, concerned with the design, monitoring, and construction of ecosystems. According to Mitsch (1996) "the design of sustainable ecosystems intends to integrate human society with its natural environment for the benefit of both".

### Overview

Ecological engineering emerged as a new idea in the early 1960s, but its definition has taken several decades to refine, its implementation is still undergoing adjustment, and its broader recognition as a new paradigm is relatively recent. Ecological engineering was introduced by Howard Odum and others as utilizing natural energy sources as the predominant input to manipulate and control environmental systems.

Mitsch and Jorgensen wrote that ecological engineering is designing societal services such that they benefit society and nature, and later noted the design should be systems based, sustainable, and integrate society with its natural environment. Odum emphasized that self-organizational properties were a central feature to ecological engineering.

Mitsch and Jørgensen were the first to define ecological engineering and provide ecological engineering principles. Later they refined the definition and increased the number of principles. They defined and characterized ecological engineering in a 1989 book and clarified it further in their 2004 book. They suggest the goal of ecological engineering is:

1. the restoration of ecosystems that have been substantially disturbed by human activities such as environmental pollution or land disturbance, and
2. the development of new sustainable ecosystems that have both human and ecological values.

They summarized the five concepts key to ecological engineering as:

1. it is based on the self-designing capacity of ecosystems,
2. it can be a field test of ecological theory,
3. it relies on integrated system approaches,
4. it conserves non-renewable energy, and
5. it supports biological conservation.

Bergen et al. defined ecological engineering as:

1. utilizing ecological science and theory,
2. applying to all types of ecosystems,
3. adapting engineering design methods, and
4. acknowledging a guiding value system.

Barrett (1999) offers a more literal definition of the term: "the design, construction, operation and management (that is, engineering) of landscape/aquatic structures and associated plant and animal communities (that is, ecosystems) to benefit humanity and, often, nature." Barrett continues: "other terms with equivalent or similar meanings include ecotechnology and two terms most often used in the erosion control field: soil bioengineering and biotechnical engineering. However, ecoengineering should not be confused with 'biotechnology' when describing genetic engineering at the cellular level, or 'bioengineering' meaning construction of artificial body parts."

This engineering discipline combines basic and applied science from engineering, ecology, economics, and natural sciences for the restoration and construction of aquatic and terrestrial ecosystems. The field of ecological engineering is increasing in breadth and depth as more opportunities to design and use ecosystems as interfaces between technology and environment are explored.

Implementation of ecological engineering has focused on the creation or restoration of ecosystems, from degraded wetlands to multi-celled tubs and greenhouses that integrate microbial, fish, and plant services to process human wastewater into products such as fertilizers, flowers, and drinking water.

Potential applications of ecological engineering in cities have included the field of landscape architecture, urban planning, and urban horticulture, which can be synthesized into urban stormwater management. Potential applications of ecological engineering in rural landscapes have included wetland treatment and community reforestation through traditional ecological knowledge.

Recent lifestyle and habitat planning examples include the permaculture movement.

## Design Guidelines

Ecological engineering design will follow a cycle similar to engineering design - problem formulation (goal), problem analysis (constraints), alternative solutions search, decision among alternatives, and specification of a complete solution. Elements that distinguish ecological engineering design are elaborated by many authors, however a singular approach is still absent. Typically, the design goal involves protecting an at-risk ecosystem, restoring a degraded ecosystem, or creating a new sustainable ecosystem to satisfy needs of nature and society.

A temporal framework is provided by Matlock et al., stating the design solutions are considered in ecological time. In selecting between alternatives, the design should incorporate ecological economics in design evaluation and acknowledge a guiding value system which promotes biological conservation.

- applying to all types of ecosystems,
- adapting engineering design methods, and
- Design steps should be based on utilizing ecological science and theory,
- the self-designing capacity of ecosystems;
- accept the adaptive management theory of learning from mistakes as the design will field test ecological theory;
- utilize integrated system approaches; and
- conserve non-renewable energy.

## Academic Curriculum

An academic curriculum has been proposed for ecological engineering, and institutions around the world are starting programs. Key elements of this curriculum are:

- quantitative ecology,
- systems ecology,
- restoration ecology,
- ecological modeling,
- ecological engineering,
- economics of ecological engineering, and
- technical electives.

Complementing this set of courses are prerequisites courses in physical, biological, and chemical subject areas, and integrated design experiences. According to Matlock et al., the design should identify constraints, characterize solutions in ecological time, and incorporate ecological economics in design evaluation. Economics of ecological engineering has been demonstrated using energy principles for a wetland., and using nutrient valuation for a dairy farm

## Literature


- Howard T. Odum (1963), "Man and Ecosystem" Proceedings, Lockwood Conference on the Suburban Forest and Ecology, in: *Bulletin Connecticut Agric. Station.*
- *W.J. Mitsch and S.E. Jørgensen (1989). Ecological Engineering: An Introduction to Ecotechnology. John Wiley and Sons.*
- W.J. Mitsch (1993), *Ecological engineering—"a cooperative role with the planetary life–support systems. Environmental Science & Technology* 27:438-445.
- *K. R. Barrett (1999). "Ecological engineering in water resources: The benefits of collaborating with nature". Water International. 24: 182–188. doi:10.1080/02508069908692160.*
- *P.C. Kangas (2004). Ecological Engineering: Principles and Practice. Lewis Publishers, CRC Press. ISBN 978-1566705998.*
- *W.J. Mitsch and S.E. Jørgensen (2004). Ecological Engineering and Ecosystem Restoration. John Wiley and Sons. ISBN 978-0471332640.*
- H.D. van Bohemen (2004), *Ecological Engineering and Civil Engineering works*, Doctoral thesis TU Delft, The Netherlands.

## References


- Eccleston, Charles; Doub, J. Peyton (2012). Preparing NEPA Environmental Assessments: A User's Guide to Best Professional Practices. CRC Press. ISBN 9781439808825.
- Patel, Kant; Rushefsky, Mark E.; McFarlane, Deborah R. (2005). The Politics of Public Health in the United States. M.E. Sharpe. p. 91. ISBN 0-7656-1135-X.
- White, Kerr L. (1991). Healing the schism: Epidemiology, medicine, and the public's health. New York: Springer-Verlag. ISBN 0-387-97574-8.
- Darnell, Regna (2008). Histories of anthropology annual. University of Nebraska Press. p. 36. ISBN 0-8032-6664-2.
- Amanda J. Thomas (2010). The Lambeth cholera outbreak of 1848-1849: the setting, causes, course and aftermath of an epidemic in London. McFarland. pp. 55–6. ISBN 978-0-7864-3989-8. Retrieved 5 April 2012.
- Margaret Stacey (1 June 2004). The Sociology of Health and Healing. Taylor and Francis. p. 69. ISBN 978-0-203-38004-8. Retrieved 5 April 2012.
- Samuel Edward Finer (1952). The Life and Times of Sir Edwin Chadwick. Methuen. pp. 424–5. ISBN 978-0-416-17350-5.
- Mooney, Graham (2015). Intrusive Interventions: Public Health, Domestic Space, and Infectious Disease Surveillance in England, 1840-1914. Rochester, NY: University of Rochester Press. ISBN 9781580465274.
- Vinten-Johansen, Peter, et al. (2003). Cholera, Chloroform, and the Science of Medicine: A Life of John Snow. Oxford University Press. ISBN 0-19-513544-X
- Johnson, Steven (2006). The Ghost Map: The Story of London's Most Terrifying Epidemic – and How it Changed Science, Cities and the Modern World. Riverhead Books. ISBN 1-59448-925-4
- Edward Marriott (1966) in "Plague. A Story of Science, Rivalry and the Scourge That Won't Go Away" ISBN 978-1-4223-5652-4
- Pierce J.R., J, Writer. 2005. Yellow Jack: How Yellow Fever Ravaged America and Walter Reed Discovered its Deadly Secrets. John Wiley and Sons. ISBN 0-471-47261-1
- Tchobanoglous, G., Burton, F.L., and Stensel, H.D. (2003). Wastewater Engineering (Treatment Disposal Reuse) / Metcalf & Eddy, Inc. (4th ed.). McGraw-Hill Book Company. ISBN 0-07-041878-0.
- Rosen (Ed.), Erwin M. (1975). The Peterson automotive troubleshooting & repair manual. Grosset & Dunlap, Inc. ISBN 978-0-448-11946-5.
- Morgan, P. (2007). Toilets That Make Compost - Low-cost, sanitary toilets that produce valuable compost for crops in an African context. Stockholm Environment Institute, ISBN 978-9-197-60222-8
- Brown, AD (2003). Feed or feedback: agriculture, population dynamics and the state the planet. International Books. Utrecht, The Netherlands. ISBN 90 5727 048X
- Lüthi, C.; et al. (2011). Sustainable sanitation in cities : a framework for action. Rijswijk: Papiroz Publ. House. p. 38. ISBN 978-90-814088-4-4. Retrieved 19 November 2015.
- Eds.; Simpson-Hébert, co-authors: Uno Winblad, Mayling (2004). Ecological sanitation (2., rev. and enlarged ed.). Stockholm: Stockholm Environment Institute. p. 67. ISBN 9188714985.
- Closing the loop ecological sanitation for food security. [Stockholm]: SIDA. 2001. ISBN 9158689354. |first1= missing |last1= in Authors list (help)

- Rosemarin, Arno; McConville, Jennifer; Flores, Amparo; Qiang, Zhu (2012). The challenges of urban ecological sanitation : lessons from the Erdos eco-town project. Practical Action Publishers. p. 116. ISBN 1853397687.
- N.J Todd & J. Todd (1994). From Eco-Cities to Living Machines: Principles of Ecological Design. North Atlantic Books. ISBN 978-1556431500.
- "Sanitation innovator named 2013 Stockholm Water Prize Laureate". Stockholm International Water Institute (SIWI). 2013. Retrieved 13 July 2015.
- ""Water and Food Security for Latin America", Fortaleza". Sustainable Sanitation Alliance. SuSanA. 2007. Retrieved 13 July 2015.
- "2015 - United Nations sustainable development agenda". United Nations Sustainable Development. Retrieved 2015-11-25.
- "Hunger and food security - United Nations Sustainable Development". United Nations Sustainable Development. Retrieved 2015-11-25.

# 3

# Management of Environment: Issues and Concerns

There are several issues to be considered by environmental engineers, some of which are resource depletion, climate change, global warming, environmental degradation, waste management etc. This chapter presents all these concerns and educates the reader about the local and global scenarios that impact environmental engineering.

## Environmental Issue

Environmental issues are harmful effects of human activity on the biophysical environment. Environmental protection is a practice of protecting the natural environment on individual, organizational or governmental levels, for the benefit of both the environment and humans. Environmentalism, a social and environmental movement, addresses environmental issues through advocacy, education and activism.

Water pollution is an environmental issue that affects many water bodies. This photograph shows foam on the New River as it enters the United States from Mexico.

The carbon dioxide equivalent of greenhouse gases (GHG) in the atmosphere has already exceeded 400 parts per million (NOAA) (with total "long-term" GHG exceeding 455 parts per million). (Intergovernmental Panel on Climate Change Report) This level is considered a tipping point. "The amount of greenhouse gas in the atmosphere is already above the threshold that can potentially cause dangerous climate change. We are already at risk of many areas of pollution...It's not next year or next decade, it's now."

Report from the UN Office for the Coordination of Humanitarian Affairs (OCHA):

"Climate disasters are on the rise. Around 70 percent of disasters are now climate related – up from around 50 percent from two decades ago.

These disasters take a heavier human toll and come with a higher price tag. In the last decade, 2.4 billion people were affected by climate related disasters, compared to 1.7 billion in the previous decade. The cost of responding to disasters has risen tenfold between 1992 and 2008. Destructive sudden heavy rains, intense tropical storms, repeated flooding and droughts are likely to increase, as will the vulnerability of local communities in the absence of strong concerted action." (OCHA) "Climate change is not just a distant future threat. It is the main driver behind rising humanitarian needs and we are seeing its impact. The number of people affected and the damages inflicted by extreme weather has been unprecedented."

Environment destruction caused by humans is a global problem, and this is a problem that is on going every day. By year 2050, the global human population is expected to grow by 2 billion people, thereby reaching a level of 9.6 billion people (Living Blue Planet 24). The human effects on Earth can be seen in many different ways. A main one is the temperature rise, and according to the report "Our Changing Climate", the global warming that has been going on for the past 50 years is primarily due to human activities (Walsh, et al. 20). Since 1895, the U.S. average temperature has increased from 1.3 °F to 1.9 °F, with most of the increase taken place since around year 1970 (Walsh, et al. 20).

## Types

Major current environmental issues may include climate change, pollution, environmental degradation, and resource depletion etc. The conservation movement lobbies for protection of endangered species and protection of any ecologically valuable natural areas, genetically modified foods and global warming.

## Scientific Grounding

The level of understanding of Earth has increased markedly in recent times through science especially with the application of the scientific method. Environmental science is now a multi-disciplinary academic study taught and researched at many universities. This is used as a basis for addressing environmental issues.

Large amounts of data have been gathered and these are collated into reports, of which a common type is the State of the Environment publications. A recent major report was the Millennium Ecosystem Assessment, with input from 1200 scientists and released in 2005, which showed the high level of impact that humans are having on ecosystem services.

## Organizations

Environmental issues are addressed at a regional, national or international level by government organizations.

The largest international agency, set up in 1972, is the United Nations Environment Programme. The International Union for Conservation of Nature brings together 83 states, 108 government agencies, 766 Non-governmental organizations and 81 international organizations and about 10,000 experts and scientists from countries around the world. International non-governmental organizations include Greenpeace, Friends of the Earth and World Wide Fund for Nature. Governments enact environmental policy and enforce environmental law and this is done to differing degrees around the world.

## Solutions

Sustainability is the key to prevent or reduce the effect of environmental issues. There is now clear scientific evidence that humanity is living unsustainably, and that an unprecedented collective effort is needed to return human use of natural resources to within sustainable limits. For humans to live sustainably, the Earth's resources must be used at a rate at which they can be replenished.

Concerns for the environment have prompted the formation of Green parties, political parties that seek to address environmental issues. Initially these were formed in Australia, New Zealand and Germany but are now present in many other countries.

## Film and Television

There are an increasing number of films being produced on environmental issues, especially on climate change and global warming. Al Gore's 2006 film *An Inconvenient Truth* gained commercial success and a high media profile.

# Climate Change

Climate change is a change in the statistical distribution of weather patterns when that change lasts for an extended period of time (i.e., decades to millions of years). Climate change may refer to a change in average weather conditions, or in the time variation of weather around longer-term average conditions (i.e., more or fewer extreme weather events). Climate change is caused by factors such as biotic processes, variations in solar radiation received by Earth, plate tectonics, and volcanic eruptions. Certain human activities have also been identified as significant causes of recent climate change, often referred to as *global warming*.

Scientists actively work to understand past and future climate by using observations and theoretical models. A climate record—extending deep into the Earth's past—has been assembled, and continues to be built up, based on geological evidence from borehole temperature profiles, cores removed from deep accumulations of ice, floral and faunal records, glacial and periglacial processes, stable-isotope and other analyses of sediment layers, and records of past sea levels. More recent data are provided by the instrumental record. General circulation models, based on the physical sciences, are often used in theoretical approaches to match past climate data, make future projections, and link causes and effects in climate change.

## Terminology

The most general definition of *climate change* is a change in the statistical properties (principally its mean and spread) of the climate system when considered over long periods of time, regardless of cause. Accordingly, fluctuations over periods shorter than a few decades, such as El Niño, do not represent climate change.

The term sometimes is used to refer specifically to climate change caused by human activity, as opposed to changes in climate that may have resulted as part of Earth's natural processes. In this sense, especially in the context of environmental policy, the term *climate change* has become synonymous with *anthropogenic global warming*. Within scientific journals, *global warming* refers to surface temperature increases while *climate change* includes global warming and everything else that increasing greenhouse gas levels affect.

## Climatic Change Versus Climate Change

In 1966, the World Meteorological Organization (WMO) proposed the term climatic change to encompass all forms of climatic variability on time-scales longer than 10 years, whether the cause was natural or anthropogenic. Change was a given and climatic was used as an adjective to describe this kind of change (as opposed to political or economic change). When it was realized that human activities had a potential to drastically alter the climate, the term climate change replaced climatic change as the dominant term to reflect an anthropogenic cause. Climate change was incorporated in the title of the Intergovernmental Panel on Climate Change (IPCC) and the UN Framework Convention on Climate Change (UNFCCC). Climate change, used as a noun, became an issue rather than the technical description of changing weather.

## Causes

On the broadest scale, the rate at which energy is received from the Sun and the rate at which it is lost to space determine the equilibrium temperature and climate of Earth. This energy is distributed around the globe by winds, ocean currents, and other mechanisms to affect the climates of different regions.

Factors that can shape climate are called climate forcings or "forcing mechanisms".

These include processes such as variations in solar radiation, variations in the Earth's orbit, variations in the albedo or reflectivity of the continents and oceans, mountain-building and continental drift and changes in greenhouse gas concentrations. There are a variety of climate change feedbacks that can either amplify or diminish the initial forcing. Some parts of the climate system, such as the oceans and ice caps, respond more slowly in reaction to climate forcings, while others respond more quickly. There are also key threshold factors which when exceeded can produce rapid change.

Forcing mechanisms can be either "internal" or "external". Internal forcing mechanisms are natural processes within the climate system itself (e.g., the thermohaline circulation). External forcing mechanisms can be either natural (e.g., changes in solar output) or anthropogenic (e.g., increased emissions of greenhouse gases).

Whether the initial forcing mechanism is internal or external, the response of the climate system might be fast (e.g., a sudden cooling due to airborne volcanic ash reflecting sunlight), slow (e.g. thermal expansion of warming ocean water), or a combination (e.g., sudden loss of albedo in the arctic ocean as sea ice melts, followed by more gradual thermal expansion of the water). Therefore, the climate system can respond abruptly, but the full response to forcing mechanisms might not be fully developed for centuries or even longer.

## Internal Forcing Mechanisms

Scientists generally define the five components of earth's climate system to include atmosphere, hydrosphere, cryosphere, lithosphere (restricted to the surface soils, rocks, and sediments), and biosphere. Natural changes in the climate system ("internal forcings") result in internal "climate variability". Examples include the type and distribution of species, and changes in ocean currents.

## Ocean Variability

The ocean is a fundamental part of the climate system, some changes in it occurring at longer timescales than in the atmosphere, as it has hundreds of times more mass and thus very high thermal inertia, with effects such as the ocean depths still lagging today in temperature adjustment from effects of the Little Ice Age of past centuries).

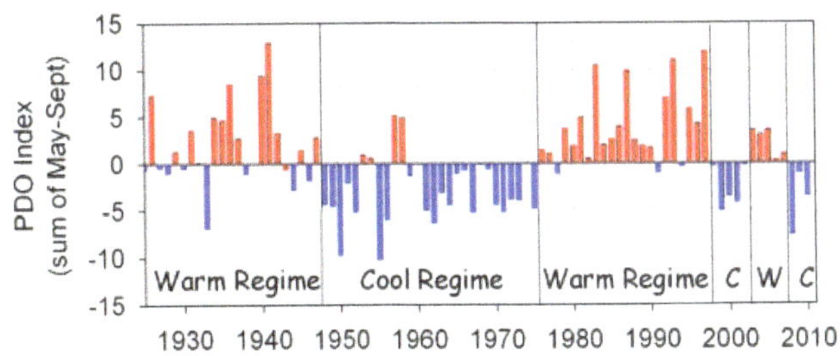


Pacific Decadal Oscillation 1925 to 2010

Short-term fluctuations (years to a few decades) such as the El Niño-Southern Oscillation, the Pacific decadal oscillation, the North Atlantic oscillation, and the Arctic oscillation, represent climate variability rather than climate change. On longer time-scales, alterations to ocean processes such as thermohaline circulation play a key role in redistributing heat by carrying out a very slow and extremely deep movement of water and the long-term redistribution of heat in the world's oceans.

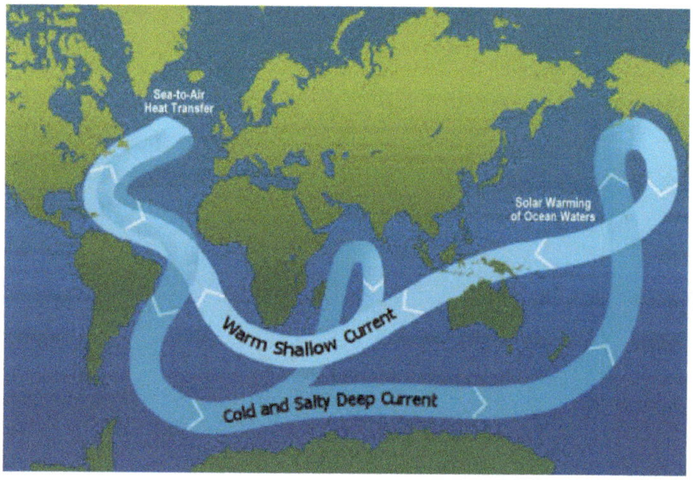


A schematic of modern thermohaline circulation. Tens of millions of years ago, continental-plate movement formed a land-free gap around Antarctica, allowing the formation of the ACC, which keeps warm waters away from Antarctica.

## Life

Life affects climate through its role in the carbon and water cycles and through such mechanisms as albedo, evapotranspiration, cloud formation, and weathering. Examples of how life may have affected past climate include:

- glaciation 2.3 billion years ago triggered by the evolution of oxygenic photosynthesis, which depleted the atmosphere of the greenhouse gas carbon dioxide and introduced free oxygen.
- another glaciation 300 million years ago ushered in by long-term burial of decomposition-resistant detritus of vascular land-plants (creating a carbon sink and forming coal)
- termination of the Paleocene-Eocene Thermal Maximum 55 million years ago by flourishing marine phytoplankton
- reversal of global warming 49 million years ago by 800,000 years of arctic azolla blooms
- global cooling over the past 40 million years driven by the expansion of grass-grazer ecosystems

## External Forcing Mechanisms

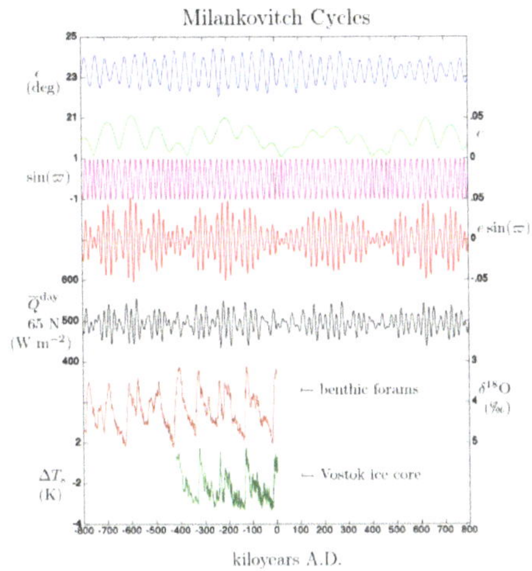


Milankovitch cycles from 800,000 years ago in the past to 800,000 years in the future.

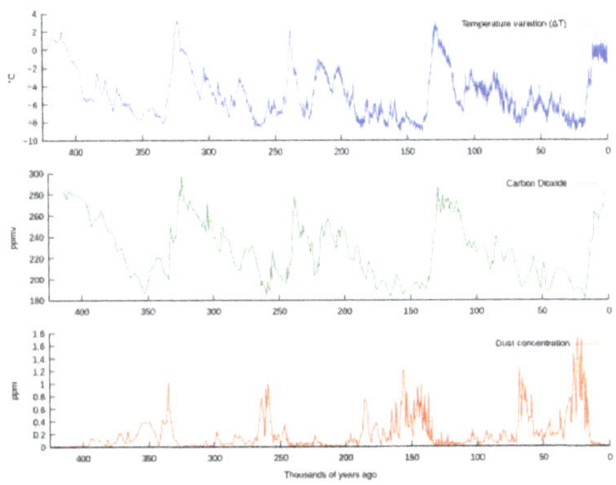


Variations in $CO_2$, temperature and dust from the Vostok ice core over the last 450,000 years

## Orbital Variations

Slight variations in Earth's orbit lead to changes in the seasonal distribution of sunlight reaching the Earth's surface and how it is distributed across the globe. There is very little change to the area-averaged annually averaged sunshine; but there can be strong changes in the geographical and seasonal distribution. The three types of orbital variations are variations in Earth's eccentricity, changes in the tilt angle of Earth's axis of rotation, and precession of Earth's axis. Combined together, these produce Milankovitch cycles which have a large impact on climate and are notable for their correlation to glacial and interglacial periods, their correlation with the advance and retreat of the Sahara, and for their appearance in the stratigraphic record.

The IPCC notes that Milankovitch cycles drove the ice age cycles, $CO_2$ followed temperature change "with a lag of some hundreds of years," and that as a feedback amplified temperature change. The depths of the ocean have a lag time in changing temperature (thermal inertia on such scale). Upon seawater temperature change, the solubility of $CO_2$ in the oceans changed, as well as other factors impacting air-sea $CO_2$ exchange.

## Solar Output

The Sun is the predominant source of energy input to the Earth. Other sources include geothermal energy from the Earth's core, and heat from the decay of radioactive compounds. Both long- and short-term variations in solar intensity are known to affect global climate.

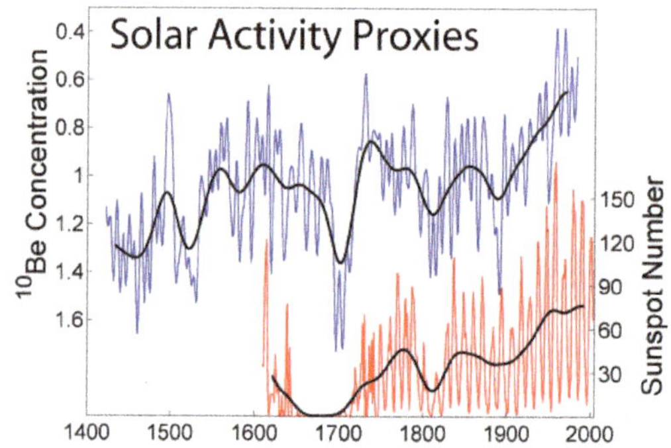


Variations in solar activity during the last several centuries based on observations of sunspots and beryllium isotopes. The period of extraordinarily few sunspots in the late 17th century was the Maunder minimum.

Three to four billion years ago, the Sun emitted only 70% as much power as it does today. If the atmospheric composition had been the same as today, liquid water should not have existed on Earth. However, there is evidence for the presence of water on the early Earth, in the Hadean and Archean eons, leading to what is known as the faint young Sun paradox. Hypothesized solutions to this paradox include a vastly different atmosphere, with much higher concentrations of greenhouse gases than currently exist. Over the following approximately 4 billion years, the energy output of the Sun increased and atmospheric composition changed. The Great Oxygenation Event – oxygenation of the atmosphere around 2.4 billion years ago – was the most notable alteration. Over the next five billion years, the Sun's ultimate death as it becomes a red giant and then a white dwarf will have large effects on climate, with the red giant phase possibly ending any life on Earth that survives until that time.

Solar output also varies on shorter time scales, including the 11-year solar cycle and longer-term modulations. Solar intensity variations possibly as a result of the Wolf, Spörer and Maunder Minimum are considered to have been influential in triggering the Little Ice Age, and some of the warming observed from 1900 to 1950. The cyclical nature of the Sun's energy output is not yet fully understood; it differs from the very slow change that is happening within the Sun as it ages and evolves. Research indicates that solar variability has had effects including the Maunder minimum from 1645 to 1715 A.D., part of the Little Ice Age from 1550 to 1850 A.D. that was marked by relative cooling and greater glacier extent than the centuries before and afterward. Some studies point toward solar radiation increases from cyclical sunspot activity affecting global warming, and climate may be influenced by the sum of all effects (solar variation, anthropogenic radiative forcings, etc.).

Interestingly, a 2010 study *suggests*, "that the effects of solar variability on temperature throughout the atmosphere may be contrary to current expectations."

In an Aug 2011 Press Release, CERN announced the publication in the Nature journal the initial results from its CLOUD experiment. The results indicate that ionisation from cosmic rays significantly enhances aerosol formation in the presence of sulfuric acid and water, but in the lower atmosphere where ammonia is also required, this is insufficient to account for aerosol formation and additional trace vapours must be involved. The next step is to find more about these trace vapours, including whether they are of natural or human origin.

## Volcanism

The eruptions considered to be large enough to affect the Earth's climate on a scale of more than 1 year are the ones that inject over 100,000 tons of $SO_2$ into the stratosphere. This is due to the optical properties of $SO_2$ and sulfate aerosols, which strongly absorb or scatter solar radiation, creating a global layer of sulfuric acid haze. On average, such eruptions occur several times per century, and cause cooling (by partially blocking the transmission of solar radiation to the Earth's surface) for a period of a few years.

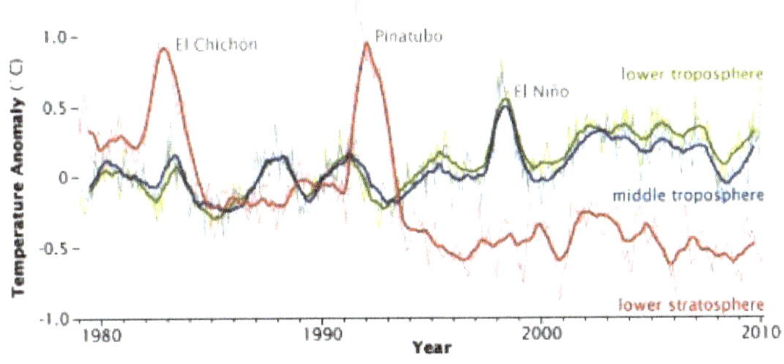


In atmospheric temperature from 1979 to 2010, determined by MSU NASA satellites, effects appear from aerosols released by major volcanic eruptions (El Chichón and Pinatubo). El Niño is a separate event, from ocean variability.

The eruption of Mount Pinatubo in 1991, the second largest terrestrial eruption of the 20th century, affected the climate substantially, subsequently global temperatures decreased by about 0.5 °C (0.9 °F) for up to three years. Thus, the cooling over large parts of the Earth reduced surface temperatures in 1991-93, the equivalent to a reduction in net radiation of 4 watts per square meter. The Mount Tambora eruption in 1815 caused the Year Without a Summer. Much larger eruptions, known as large igneous provinces, occur only a few times every fifty - hundred million years - through flood basalt, and caused in Earth past global warming and mass extinctions.

Small eruptions, with injections of less than 0.1 Mt of sulfur dioxide into the stratosphere, impact the atmosphere only subtly, as temperature changes are comparable with natural variability. However, because smaller eruptions occur at a much higher frequency, they too have a significant impact on Earth's atmosphere.

Seismic monitoring maps current and future trends in volcanic activities, and tries to develop early warning systems. In climate modelling the aim is to study the physical mechanisms and feedbacks of volcanic forcing.

Volcanoes are also part of the extended carbon cycle. Over very long (geological) time periods, they release carbon dioxide from the Earth's crust and mantle, counteracting the uptake by sedimentary rocks and other geological carbon dioxide sinks. The US Geological Survey estimates are that volcanic emissions are at a much lower level than the effects of current human activities, which generate 100–300 times the amount of carbon dioxide emitted by volcanoes. A review of published studies indicates that annual volcanic emissions of carbon dioxide, including amounts released from mid-ocean ridges, volcanic arcs, and hot spot volcanoes, are only the equivalent of 3 to 5 days of human caused output. The annual amount put out by human activities may be greater than the amount released by supererruptions, the most recent of which was the Toba eruption in Indonesia 74,000 years ago.

Although volcanoes are technically part of the lithosphere, which itself is part of the climate system, the IPCC explicitly defines volcanism as an external forcing agent.

## Plate Tectonics

Over the course of millions of years, the motion of tectonic plates reconfigures global land and ocean areas and generates topography. This can affect both global and local patterns of climate and atmosphere-ocean circulation.

The position of the continents determines the geometry of the oceans and therefore influences patterns of ocean circulation. The locations of the seas are important in controlling the transfer of heat and moisture across the globe, and therefore, in determining global climate. A recent example of tectonic control on ocean circulation is the formation of the Isthmus of Panama about 5 million years ago, which shut off direct mixing between the Atlantic and Pacific Oceans. This strongly affected the ocean dynamics of what is now the Gulf Stream and may have led to Northern Hemisphere ice cover. During the Carboniferous period, about 300 to 360 million years ago, plate tectonics may have triggered large-scale storage of carbon and increased glaciation. Geologic evidence points to a "megamonsoonal" circulation pattern during the time of the supercontinent Pangaea, and climate modeling suggests that the existence of the supercontinent was conducive to the establishment of monsoons.

The size of continents is also important. Because of the stabilizing effect of the oceans on temperature, yearly temperature variations are generally lower in coastal areas than they are inland. A larger supercontinent will therefore have more area in which climate is strongly seasonal than will several smaller continents or islands.

## Human Influences

In the context of climate variation, anthropogenic factors are human activities which affect the climate. The scientific consensus on climate change is "that climate is changing and that these changes are in large part caused by human activities," and it "is largely irreversible."

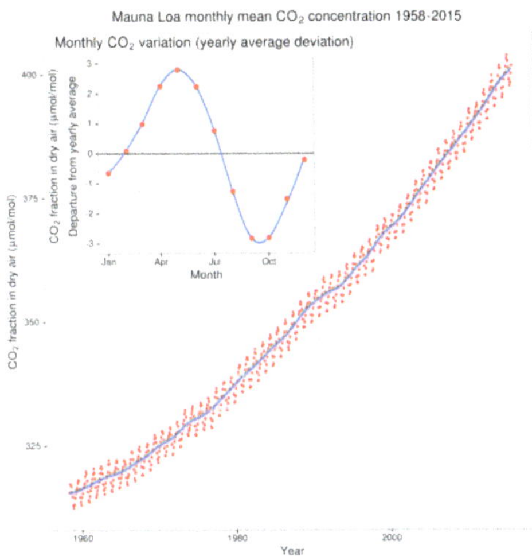


Increase in atmospheric CO2 levels

"Science has made enormous inroads in understanding climate change and its causes, and is beginning to help develop a strong understanding of current and potential impacts that will affect people today and in coming decades. This understanding is crucial because it allows decision makers to place climate change in the context of other large challenges facing the nation and the world. There are still some uncertainties, and there always will be in understanding a complex system like Earth's climate. Nevertheless, there is a strong, credible body of evidence, based on multiple lines of research, documenting that climate is changing and that these changes are in large part caused by human activities. While much remains to be learned, the core phenomenon, scientific questions, and hypotheses have been examined thoroughly and have stood firm in the face of serious scientific debate and careful evaluation of alternative explanations."

*— United States National Research Council, Advancing the Science of Climate Change*

Of most concern in these anthropogenic factors is the increase in $CO_2$ levels due to emissions from fossil fuel combustion, followed by aerosols (particulate matter in the atmosphere) and the $CO_2$ released by cement manufacture. Other factors, including land use, ozone depletion, animal agriculture and deforestation, are also of concern in the roles they play – both separately and in conjunction with other factors – in affecting climate, microclimate, and measures of climate variables.

## Physical Evidence

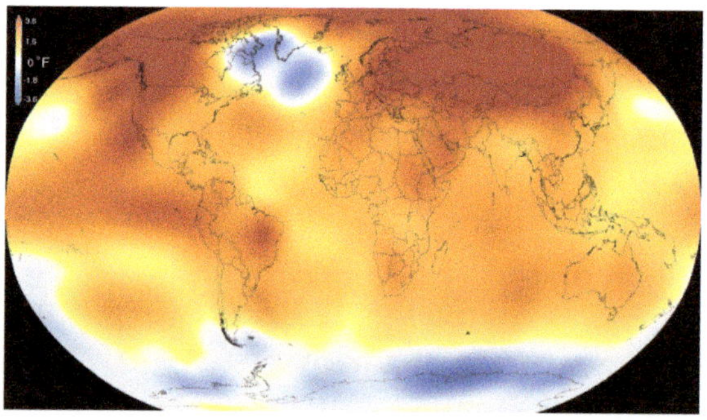


2015 – Warmest Global Year on Record (since 1880) – Colors indicate temperature anomalies (NASA/NOAA; 20 January 2016).

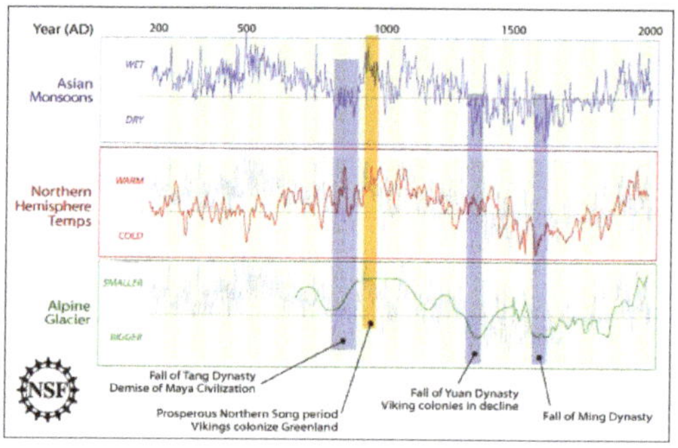


Comparisons between Asian Monsoons from 200 AD to 2000 AD (staying in the background on other plots), Northern Hemisphere temperature, Alpine glacier extent (vertically inverted as marked), and human history as noted by the U.S. NSF.

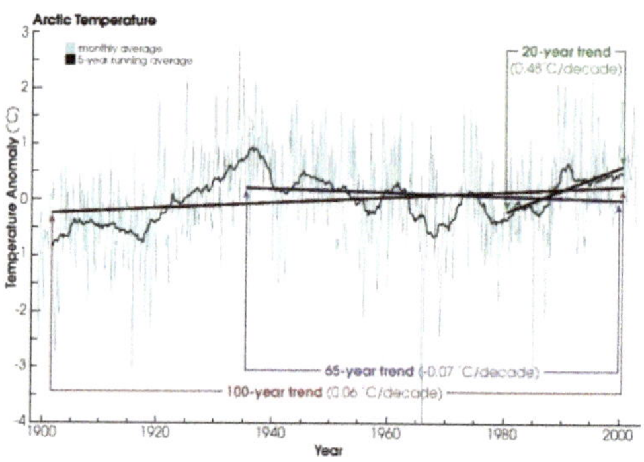


Arctic temperature anomalies over a 100-year period as estimated by NASA. Typical high monthly variance can be seen, while longer-term averages highlight trends.

Evidence for climatic change is taken from a variety of sources that can be used to reconstruct past climates. Reasonably complete global records of surface temperature are available beginning from the mid-late 19th century. For earlier periods, most of the evidence is indirect—climatic changes are inferred from changes in proxies, indicators that reflect climate, such as vegetation, ice cores, dendrochronology, sea level change, and glacial geology.

## Temperature Measurements and Proxies

The instrumental temperature record from surface stations was supplemented by radiosonde balloons, extensive atmospheric monitoring by the mid-20th century, and, from the 1970s on, with global satellite data as well. The O/O ratio in calcite and ice core samples used to deduce ocean temperature in the distant past is an example of a temperature proxy method, as are other climate metrics noted in subsequent categories.

## Historical and Archaeological Evidence

Climate change in the recent past may be detected by corresponding changes in settlement and agricultural patterns. Archaeological evidence, oral history and historical documents can offer insights into past changes in the climate. Climate change effects have been linked to the collapse of various civilizations.

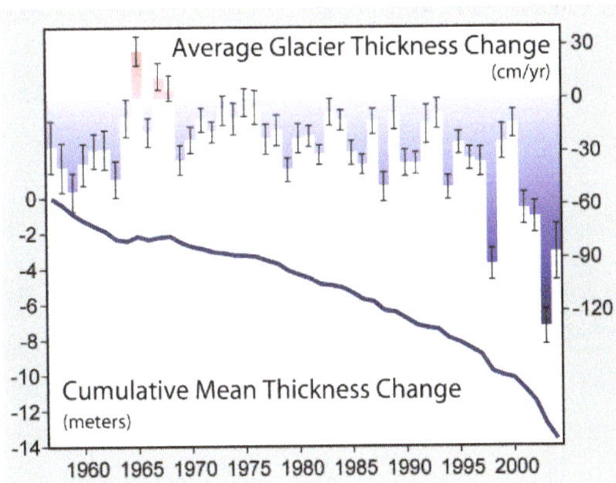


Decline in thickness of glaciers worldwide over the past half-century

## Glaciers

Glaciers are considered among the most sensitive indicators of climate change. Their size is determined by a mass balance between snow input and melt output. As temperatures warm, glaciers retreat unless snow precipitation increases to make up for the additional melt; the converse is also true.

Glaciers grow and shrink due both to natural variability and external forcings. Variability in temperature, precipitation, and englacial and subglacial hydrology can strongly determine the evolution of a glacier in a particular season. Therefore, one must average over a decadal or longer time-scale and/or over many individual glaciers to smooth out the local short-term variability and obtain a glacier history that is related to climate.

A world glacier inventory has been compiled since the 1970s, initially based mainly on aerial photographs and maps but now relying more on satellites. This compilation tracks more than 100,000 glaciers covering a total area of approximately 240,000 km, and preliminary estimates indicate that the remaining ice cover is around 445,000 km. The World Glacier Monitoring Service collects data annually on glacier retreat and glacier mass balance. From this data, glaciers worldwide have been found to be shrinking significantly, with strong glacier retreats in the 1940s, stable or growing conditions during the 1920s and 1970s, and again retreating from the mid-1980s to present.

The most significant climate processes since the middle to late Pliocene (approximately 3 million years ago) are the glacial and interglacial cycles. The present interglacial period (the Holocene) has lasted about 11,700 years. Shaped by orbital variations, responses such as the rise and fall of continental ice sheets and significant sea-level changes helped create the climate. Other changes, including Heinrich events, Dansgaard–Oeschger events and the Younger Dryas, however, illustrate how glacial variations may also influence climate without the orbital forcing.

Glaciers leave behind moraines that contain a wealth of material—including organic matter, quartz, and potassium that may be dated—recording the periods in which a glacier advanced and retreated. Similarly, by tephrochronological techniques, the lack of glacier cover can be identified by the presence of soil or volcanic tephra horizons whose date of deposit may also be ascertained.

## Arctic Sea Ice Loss

The decline in Arctic sea ice, both in extent and thickness, over the last several decades is further evidence for rapid climate change. Sea ice is frozen seawater that floats on the ocean surface. It covers millions of square miles in the polar regions, varying with the seasons. In the Arctic, some sea ice remains year after year, whereas almost all Southern Ocean or Antarctic sea ice melts away and reforms annually. Satellite observations show that Arctic sea ice is now declining at a rate of 13.3 percent per decade, relative to the 1981 to 2010 average.

This video summarizes how climate change, associated with increased carbon dioxide levels, has affected plant growth.

## Vegetation

A change in the type, distribution and coverage of vegetation may occur given a change in the climate. Some changes in climate may result in increased precipitation and warmth, resulting in improved plant growth and the subsequent sequestration of airborne $CO_2$. A gradual increase in warmth in a region will lead to earlier flowering and fruiting times, driving a change in the timing of life cycles of dependent organisms. Conversely, cold will cause plant bio-cycles to lag. Larger, faster or more radical changes, however, may result in vegetation stress, rapid plant loss and desertification in certain circumstances. An example of this occurred during the Carboniferous Rainforest Collapse (CRC), an extinction event 300 million years ago. At this time vast rainforests covered the equatorial region of Europe and America. Climate change devastated these tropical rainforests, abruptly fragmenting the habitat into isolated 'islands' and causing the extinction of many plant and animal species.

## Pollen Analysis

Palynology is the study of contemporary and fossil palynomorphs, including pollen. Palynology is used to infer the geographical distribution of plant species, which vary under different climate conditions. Different groups of plants have pollen with distinctive shapes and surface textures, and since the outer surface of pollen is composed of a very resilient material, they resist decay. Changes in the type of pollen found in different layers of sediment in lakes, bogs, or river deltas indicate changes in plant communities. These changes are often a sign of a changing climate. As an example, palynological studies have been used to track changing vegetation patterns throughout the Quaternary glaciations and especially since the last glacial maximum.

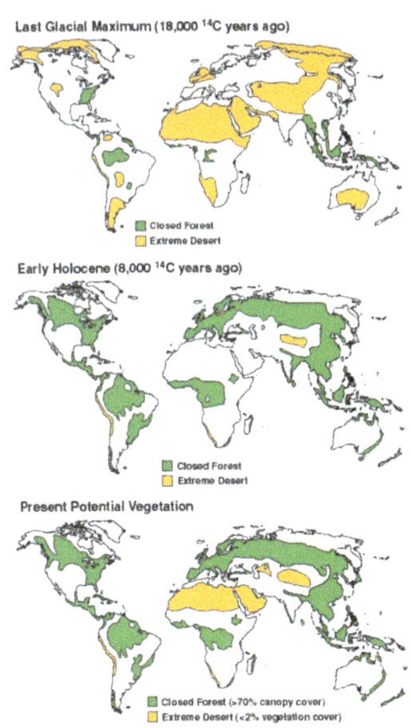


*Top:* Arid ice age climate
*Middle:* Atlantic Period, warm and wet
*Bottom:* Potential vegetation in climate now if not for human effects like agriculture.

## Cloud Cover and Precipitation

Past precipitation can be estimated in the modern era with the global network of precipitation gauges. Surface coverage over oceans and remote areas is relatively sparse, but, reducing reliance on interpolation, satellite clouds and precipitation data has been available since the 1970s. Quantification of climatological variation of precipitation in prior centuries and epochs is less complete but approximated using proxies such as marine sediments, ice cores, cave stalagmites, and tree rings. In July 2016 scientists published evidence of increased cloud cover over polar regions, as predicted by climate models.

Climatological temperatures substantially affect cloud cover and precipitation. For instance, during the Last Glacial Maximum of 18,000 years ago, thermal-driven evaporation from the oceans onto continental landmasses was low, causing large areas of extreme desert, including polar deserts (cold but with low rates of cloud cover and precipitation). In contrast, the world's climate was cloudier and wetter than today near the start of the warm Atlantic Period of 8000 years ago.

Estimated global land precipitation increased by approximately 2% over the course of the 20th century, though the calculated trend varies if different time endpoints are chosen, complicated by ENSO and other oscillations, including greater global land cloud cover precipitation in the 1950s and 1970s than the later 1980s and 1990s despite the positive trend over the century overall. Similar slight overall increase in global river runoff and in average soil moisture has been perceived.

## Dendroclimatology

Dendroclimatology is the analysis of tree ring growth patterns to determine past climate variations. Wide and thick rings indicate a fertile, well-watered growing period, whilst thin, narrow rings indicate a time of lower rainfall and less-than-ideal growing conditions.

## Ice Cores

Analysis of ice in a core drilled from an ice sheet such as the Antarctic ice sheet, can be used to show a link between temperature and global sea level variations. The air trapped in bubbles in the ice can also reveal the $CO_2$ variations of the atmosphere from the distant past, well before modern environmental influences. The study of these ice cores has been a significant indicator of the changes in $CO_2$ over many millennia, and continues to provide valuable information about the differences between ancient and modern atmospheric conditions.

## Animals

Remains of beetles are common in freshwater and land sediments. Different species of beetles tend to be found under different climatic conditions. Given the extensive lineage of beetles whose genetic makeup has not altered significantly over the millennia, knowledge of the present climatic range of the different species, and the age of the sediments in which remains are found, past climatic conditions may be inferred.

Similarly, the historical abundance of various fish species has been found to have a substantial relationship with observed climatic conditions. Changes in the primary productivity of autotrophs in the oceans can affect marine food webs.

### Sea Level Change

Global sea level change for much of the last century has generally been estimated using tide gauge measurements collated over long periods of time to give a long-term average. More recently, altimeter measurements — in combination with accurately determined satellite orbits — have provided an improved measurement of global sea level change. To measure sea levels prior to instrumental measurements, scientists have dated coral reefs that grow near the surface of the ocean, coastal sediments, marine terraces, ooids in limestones, and nearshore archaeological remains. The predominant dating methods used are uranium series and radiocarbon, with cosmogenic radionuclides being sometimes used to date terraces that have experienced relative sea level fall. In the early Pliocene, global temperatures were 1–2˚C warmer than the present temperature, yet sea level was 15–25 meters higher than today.

## Environmental Degradation

Environmental degradation is the deterioration of the environment through depletion of resources such as air, water and soil; the destruction of ecosystems and the extinction of wildlife. It is defined as any change or disturbance to the environment perceived to be deleterious or undesirable. As indicated by the I=PAT equation, environmental impact (I) or degradation is caused by the combination of an already very large and increasing human population (P), continually increasing economic growth or per capita affluence (A), and the application of resource depleting and polluting technology (T).

Eighty-plus years after the abandonment of Wallaroo Mines (Kadina, South Australia), mosses remain the only vegetation at some spots of the site's grounds

Environmental degradation is one of the ten threats officially cautioned by the High-level Panel on Threats, Challenges and Change of the United Nations. The United Nations International Strategy for Disaster Reduction defines environmental degradation as "The reduction of the capacity of the environment to meet social and ecological objectives, and needs". Environmental degradation is of many types. When natural habitats are destroyed or natural resources are depleted, the environment is degraded. Efforts to counteract this problem include environmental protection and environmental resources management.

## Water Degradation

One major component of environmental degradation is the depletion of the resource of fresh water on Earth. Approximately only 2.5% of all of the water on Earth is fresh water, with the rest being salt water. 69% of the fresh water is frozen in ice caps located on Antarctica and Greenland, so only 30% of the 2.5% of fresh water is available for consumption. Fresh water is an exceptionally important resource, since life on Earth is ultimately dependent on it. Water transports nutrients and chemicals within the biosphere to all forms of life, sustains both plants and animals, and moulds the surface of the Earth with transportation and deposition of materials.

The current top three uses of fresh water account for 95% of its consumption; approximately 85% is used for irrigation of farmland, golf courses, and parks, 6% is used for domestic purposes such as indoor bathing uses and outdoor garden and lawn use, and 4% is used for industrial purposes such as processing, washing, and cooling in manufacturing centers. It is estimated that one in three people over the entire globe are already facing water shortages, almost one-fifth of the world's population live in areas of physical water scarcity, and almost one quarter of the world's population live in a developing country that lacks the necessary infrastructure to use water from available rivers and aquifers. Water scarcity is an increasing problem due to many foreseen issues in the future, including population growth, increased urbanization, higher standards of living, and climate change.

## Climate Change and Temperature

Climate change affects the Earth's water supply in a large number of ways. It is predicted that the mean global temperature will rise in the coming years due to a number of forces affecting the climate, the amount of atmospheric $CO_2$ will rise, and both of these will influence water resources; evaporation depends strongly on temperature and moisture availability, which can ultimately affect the amount of water available to replenish groundwater supplies.

Transpiration from plants can be affected by a rise in atmospheric $CO_2$, which can decrease their use of water, but can also raise their use of water from possible increases of leaf area. Temperature increase can decrease the length of the snow season in the winter and increase the intensity of snowmelt in warmer seasons, leading to peak runoff of snowmelt earlier in the season, affecting soil moisture, flood and drought risks, and storage capacities depending on the area.

Warmer winter temperatures cause a decrease in snowpack, which can result in diminished water resources during summer. This is especially important at mid-latitudes and in mountain regions that depend on glacial runoff to replenish their river systems and groundwater supplies, making these areas increasingly vulnerable to water shortages over time; an increase in temperature will initially result in a rapid rise in water melting from glaciers in the summer, followed by a retreat in glaciers and a decrease in the melt and consequently the water supply every year as the size of these glaciers get smaller and smaller.

Thermal expansion of water and increased melting of oceanic glaciers from an increase in temperature gives way to a rise in sea level, which can affect the fresh water supply of coastal areas as well; as river mouths and deltas with higher salinity get pushed further inland, an intrusion of saltwater results in an increase of salinity in reservoirs and aquifers. Sea-level rise may also consequently be caused by a depletion of groundwater, as climate change can affect the hydrologic cycle in a number of ways. Uneven distributions of increased temperatures and increased precipitation around the globe results in water surpluses and deficits, but a global decrease in groundwater suggests a rise in sea level, even after meltwater and thermal expansion were accounted for, which can provide a positive feedback to the problems sea-level rise causes to fresh-water supply.

A rise in air temperature results in a rise in water temperature, which is also very significant in water degradation, as the water would become more susceptible to bacterial growth. An increase in water temperature can also affect ecosystems greatly because of a species' sensitivity to temperature, and also by inducing changes in a body of water's self-purification system from decreased amounts of dissolved oxygen in the water due to rises in temperature.

## Climate Change and Precipitation

A rise in global temperatures is also predicted to correlate with an increase in global precipitation, but because of increased runoff, floods, increased rates of soil erosion, and mass movement of land, a decline in water quality is probable, while water will carry more nutrients, it will also carry more contaminants. While most of the attention about climate change is directed towards global warming and greenhouse effect, some of the most severe effects of climate change are likely to be from changes in precipitation, evapotranspiration, runoff, and soil moisture. It is generally expected that, on average, global precipitation will increase, with some areas receiving increases and some decreases.

Climate models show that while some regions should expect an increase in precipitation, such as in the tropics and higher latitudes, other areas are expected to see a decrease, such as in the subtropics; this will ultimately cause a latitudinal variation in water distribution. The areas receiving more precipitation are also expected to receive this increase during their winter and actually become drier during their summer, creating even more of a variation of precipitation distribution. Naturally, the distribution of precipitation across the planet is very uneven, causing constant variations in water availability in respective locations.

Changes in precipitation affect the timing and magnitude of floods and droughts, shift runoff processes, and alter groundwater recharge rates. Vegetation patterns and growth rates will be directly affected by shifts in precipitation amount and distribution, which will in turn affect agriculture as well as natural ecosystems. Decreased precipitation will deprive areas of water, causing water tables to fall and reservoirs and wetlands, rivers, and lakes to empty, and possibly an increase in evaporation and evapotranspiration, depending on the accompanied rise in temperature. Groundwater reserves will be depleted, and the remaining water has a greater chance of being of poor quality from saline or contaminants on the land surface.

## Population Growth

The population on Earth is expanding rapidly which goes hand in hand in the degradation of the environment at large measures. The human's appetites for needs are disarranging the environments natural equilibrium. Our production industries are venting smoke and discharging chemicals that are polluting our water. The smoke that is emitted into the atmosphere holds unappealing gases such as carbon monoxide and sulfur dioxide. The high levels of pollution in the atmosphere form layers that are eventually absorbed into the atmosphere. Organic compounds such as chlorofluorocarbons (CFC's) have generated an unwanted opening in the ozone layer, which emits higher levels of ultraviolet radiation putting the globe at large threat.

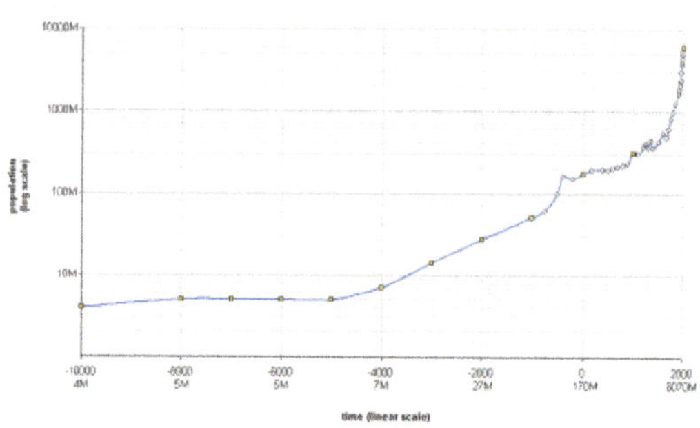


World population growth in a lin-log scale.

The available fresh water being affected by climate is also being stretched across an ever-increasing global population. It is estimated that almost a quarter of the global population is living in an area that is using more than 20% of their renewable water supply; water use will rise with population while the water is also being aggravated by decreases in streamflow and groundwater caused by climate change. Even though some areas may see an increase in freshwater supply from an uneven distribution of precipitation increase, an increased use of water supply is expected.

An increased population means increased withdrawals from the water supply for domestic, agricultural, and industrial uses, the largest of these being agriculture, believed to be the major non-climate driver of environmental change and water deterioration. The next 50 years will likely be the last period of rapid agricultural expansion, but the larger and wealthier population over this time will demand more agriculture.

Population increase over the last two decades, at least in the United States, has also been accompanied by a shift to an increase in urban areas from rural areas, which concentrates the demand for water into certain areas, and puts stress on the fresh water supply from industrial and human contaminants. Urbanization causes overcrowding and increasingly unsanitary living conditions, especially in developing countries, which in turn exposes an increasingly number of people to disease. About 79% of the world's population is in developing countries, which lack access to sanitary water and sewer systems, giving riscs to disease and deaths from contaminated water and increased numbers of disease-carrying insects.

## Agriculture

Agriculture is dependent on available soil moisture, which is directly affected by climate dynamics, with precipitation being the input in this system and various processes being the output, such as evapotranspiration, surface runoff, drainage, and percolation into groundwater. Changes in climate, especially the changes in precipitation and evapotranspiration predicted by climate models, will directly affect soil moisture, surface runoff, and groundwater recharge.

Water pollution due to dairy farming in the Wairarapa in New Zealand

In areas with decreasing precipitation as predicted by the climate models, soil moisture may be substantially reduced. With this in mind, agriculture in most areas needs irrigation already, which depletes fresh water supplies both by the physical use of the water and the degradation agriculture causes to the water. Irrigation increases salt and nutrient content in areas that would not normally be affected, and damages streams and rivers from damming and removal of water. Fertilizer enters both human and livestock waste streams that eventually enter groundwater, while nitrogen, phosphorus, and other chemicals from fertilizer can acidify both soils and water. Certain agricultural demands may increase more than others with an increasingly wealthier global population, and meat is one commodity expected to double global food demand by 2050, which directly affects the global supply of fresh water. Cows need water to drink, more if the temperature is high and humidity is low, and more if the production system the cow is in is extensive, since finding food takes more effort. Water is needed in processing of the meat, and also in the production of feed for the livestock. Manure can contaminate bodies of freshwater, and slaughterhouses, depending on how well they are managed, contribute waste such as blood, fat, hair, and other bodily contents to supplies of fresh water.

The transfer of water from agricultural to urban and suburban use raises concerns about agricultural sustainability, rural socioeconomic decline, food security, an increased carbon footprint from imported food, and decreased foreign trade balance. The depletion of fresh water, as applied to more specific and populated areas, increases fresh water scarcity among the population and also makes populations susceptible to economic, social, and political conflict in a number of ways; rising sea levels forces migration from coastal areas to other areas farther inland, pushing populations closer together breaching borders and other geographical patterns, and agricultural surpluses and deficits from the availability of water induce trade problems and economies of certain areas. Climate change is an important cause of involuntary migration and forced displacement

## Water Management

The issue of the depletion of fresh water can be met by increased efforts in water management. While water management systems are often flexible, adaptation to new hydrologic conditions may be very costly. Preventative approaches are necessary to avoid high costs of inefficiency and the need for rehabilitation of water supplies, and innovations to decrease overall demand may be important in planning water sustainability.

Water supply systems, as they exist now, were based on the assumptions of the current climate, and built to accommodate existing river flows and flood frequencies. Reservoirs are operated based on past hydrologic records, and irrigation systems on historical temperature, water availability, and crop water requirements; these may not be a reliable guide to the future. Re-examining engineering designs, operations, optimizations, and planning, as well as re-evaluating legal, technical, and economic approaches to manage water resources are very important for the future of water management in response to water degradation. Another approach is water privatization; despite its economic and cultural effects, service quality and overall quality of the water can be more easily controlled and distributed. Rationality and sustainability is appropriate, and requires limits to overexploitation and pollution, and efforts in conservation.

A stream in the town of Amlwch, Anglesey which is contaminated by acid mine drainage from the former copper mine at nearby Parys Mountain.

## Resource Depletion

Resource depletion is the consumption of a resource faster than it can be replenished. Natural resources are commonly divided between renewable resources and non-renewable resources. Use of either of these forms of resources beyond their rate of replacement is considered to be resource depletion.

Resource depletion is most commonly used in reference to farming, fishing, mining, water usage, and consumption of fossil fuels.

### Causes

- Aquifer depletion
- Habitat degradation leads to the loss of biodiversity (i.e. species and ecosystems with its ecosystem services)
- Irrigation
- Mining for fossil fuels and minerals

- Overconsumption, excessive or unnecessary use of resources
- Overpopulation
- Pollution or contamination of resources
- Slash-and-burn agricultural practices, currently occurring in many developing countries
- Soil erosion
- Technological and industrial development
- Deforestation.

## Minerals

Minerals are needed to provide food, clothing, and housing. A United States Geological Survey (USGS) study found a significant long-term trend over the 20th century for non-renewable resources such as minerals to supply a greater proportion of the raw material inputs to the non-fuel, non-food sector of the economy; an example is the greater consumption of crushed stone, sand, and gravel used in construction.

Large-scale exploitation of minerals began in the Industrial Revolution around 1760 in England and has grown rapidly ever since. Technological improvements have allowed humans to dig deeper and access lower grades and different types of ore over that time. Virtually all basic industrial metals (copper, iron, bauxite, etc.), as well as rare earth minerals, face production output limitations from time to time, because supply involves large up-front investments and is therefore slow to respond to rapid increases in demand.

Minerals projected by some to enter production decline during the next 20 years:

- Gas (2023)
- Copper (2024). Data from the United States Geological Survey (USGS) suggest that it is very unlikely that copper production will peak before 2040.
- Zinc. New developments in hydrometallurgy have transformed non-sulphide zinc deposits (largely ignored until now) into large low cost reserves.

Minerals projected by some to enter production decline during the present century:

- Aluminium (2057)
- Coal (2060)
- Iron (2068).

Such projections may change, as new discoveries are made and typically misinterpret available data on Mineral Resources and Mineral Reserves.

## Oil

Peak oil is the period when the maximum rate of global petroleum extraction is reached, after which the rate of production enters terminal decline. It relates to a long-term decline in the available supply of petroleum. This, combined with increasing demand, will significantly increase the worldwide prices of petroleum derived products. Most significant will be the availability and price of liquid fuel for transportation.

The United States Department of Energy in the Hirsch report indicates that "The peaking of world oil production presents the U. S. and the world with an unprecedented risk management problem. As peaking is approached, liquid fuel prices and price volatility will increase dramatically, and, without timely mitigation, the economic, social, and political costs will be unprecedented. Viable mitigation options exist on both the supply and demand sides, but to have substantial impact, they must be initiated more than a decade in advance of peaking."

## Deforestation

Deforestation is the clearing of forests by logging or burning of trees and plants in a forested area. As a result of deforestation, presently about one half of the forests that once covered Earth have been destroyed. It occurs for many different reasons, and it has several negative implications on the atmosphere and the quality of the land in and surrounding the forest.

## Causes

One of the main causes of deforestation is clearing forests for agricultural reasons. As the population of developing areas, especially near rainforests, increases, the need for land for farming becomes more and more important. For most people, a forest has no value when its resources aren't being used, so the incentives to deforest these areas outweigh the incentives to preserve the forests. For this reason, the economic value of the forests is very important for the developing countries.

## Environmental Impact

Because deforestation is so extensive, it has made several significant impacts on the environment, including carbon dioxide in the atmosphere, changing the water cycle, an increase in soil erosion, and a decrease in biodiversity. Deforestation is often cited as a cause of global warming. Because trees and plants remove carbon dioxide and emit oxygen into the atmosphere, the reduction of forests contribute to about 12% of anthropogenic carbon dioxide emissions. One of the most pressing issues that deforestation creates is soil erosion. The removal of trees causes higher rates of erosion, increasing risks of landslides, which is a direct threat to many people living close to deforested areas. As forests get destroyed, so does the habitat for millions of animals. It is estimated that 80% of the world's known biodiversity lives in the rainforests, and the destruction of these rainforests is accelerating extinction at an alarming rate.

## Controlling Deforestation

The United Nations and the World Bank created programs such as Reducing Emissions from Deforestation and Forest Degradation (REDD), which works especially with developing countries to use subsidies or other incentives to encourage citizens to use the forest in a more sustainable way. In addition to making sure that emissions from deforestation are kept to a minimum, an effort to educate people on sustainability and helping them to focus on the long-term risks is key to the success of these programs. The New York Declaration on Forests and its associated actions promotes reforestation, which is being encouraged in many countries in an attempt to repair the damage that deforestation has done.

## Wetlands

Wetlands are areas that are often saturated by enough surface or groundwater to sustain vegetation that is usually adapted to saturated soil conditions, such as cattails, bulrushes, red maples, wild rice, blackberries, cranberries, and peat moss. Because some varieties of wetlands are rich in minerals and nutrients and provide many of the advantages of both land and water environments they contain diverse species and possibly even form a food chain. When human activities take away resources many species are affected. An ecosystem contains many species.

Years ago people assumed wetlands were useless so it was not a large concern when they were being dug up. Many people want to use them for developing homes etc. On the other side of the argument, people believe that the wetlands are a vital source for other life forms and a part of the life cycle.

Wetlands provide services for:

1. Food and habitat
2. Improving water quality
3. Commercial fishing
4. Floodwater reduction
5. Shoreline stabilization
6. Recreation

Some loss of wetlands resulted from natural causes such as erosion, sedimentation (the buildup of soil by the settling of fine particles over a long period of time), subsidence (the sinking of land because of diminishing underground water supplies), and a rise in the sea level.

## Global Warming

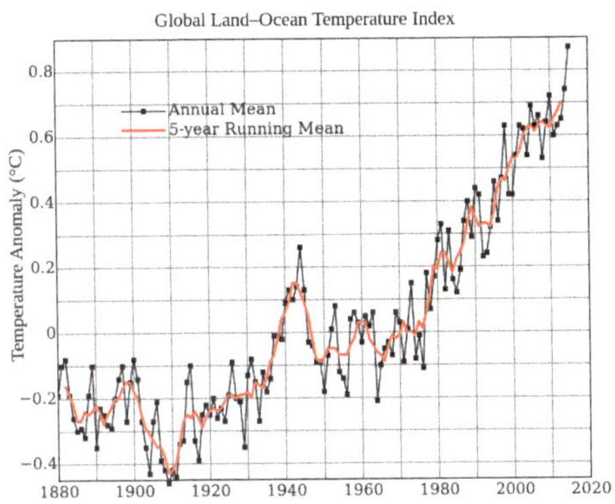


Global mean surface temperature change from 1880 to 2015, relative to the 1951–1980 mean. The black line is the annual mean and the red line is the 5-year running mean.

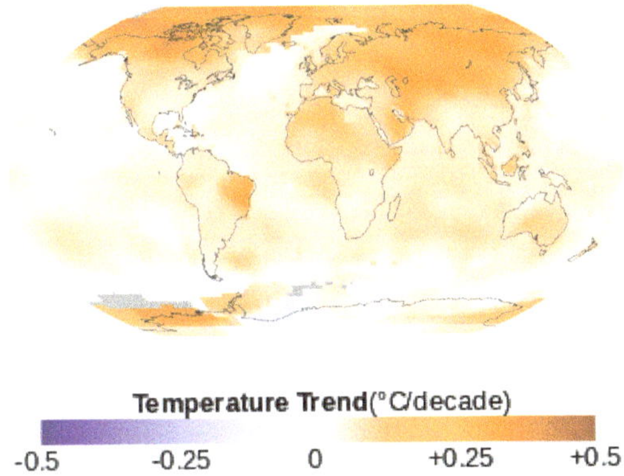


World map showing surface temperature trends (°C per decade) between 1950 and 2014.

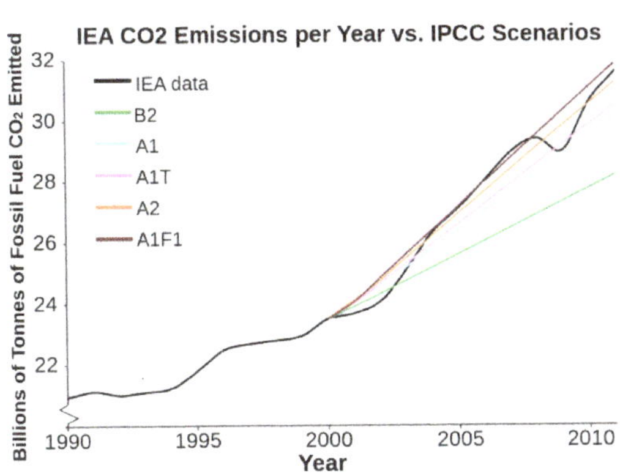


Fossil fuel related carbon dioxide ($CO_2$) emissions compared to five of the IPCC's "SRES" emissions scenarios, published in 2000. The dips are related to global recessions.

Global warming and climate change are terms for the observed century-scale rise in the average temperature of the Earth's climate system and its related effects. Multiple lines of scientific evidence show that the climate system is warming. Although the increase of near-surface atmospheric temperature is the measure of global warming often reported in the popular press, most of the additional energy stored in the climate system since 1970 has gone into the oceans. The rest has melted ice and warmed the continents and atmosphere. Many of the observed changes since the 1950s are unprecedented over tens to thousands of years.

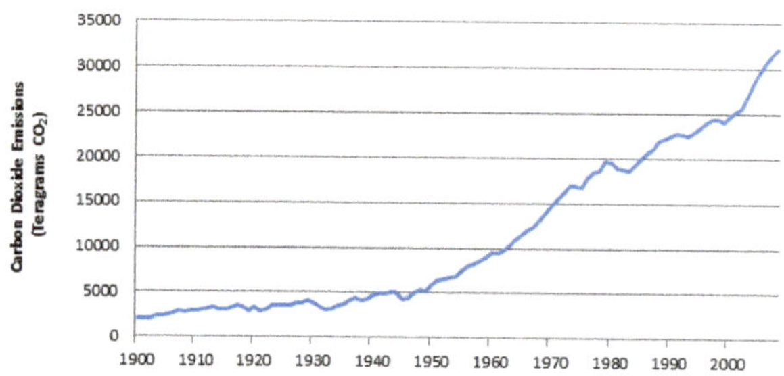


Fossil fuel related carbon dioxide emissions over the 20th century.

Scientific understanding of global warming is increasing. The Intergovernmental Panel on Climate Change (IPCC) reported in 2014 that scientists were more than 95% certain that global warming is mostly being caused by human (anthropogenic) activities, mainly increasing concentrations of greenhouse gases such as carbon dioxide ($CO_2$). Human-made carbon dioxide continues to increase above levels not seen in hundreds of thousands of years. Currently, about half of the carbon dioxide released from the burning of fossil fuels remains in the atmosphere. The rest is absorbed by vegetation and the oceans. Climate model projections summarized in the report indicated that during the 21st century the global surface temperature is likely to rise a further 0.3 to 1.7 °C (0.5 to 3.1 °F) for their lowest emissions scenario and 2.6 to 4.8 °C (4.7 to 8.6 °F) for the highest emissions scenario. These findings have been recognized by the national science academies of the major industrialized nations and are not disputed by any scientific body of national or international standing.

Future climate change and associated impacts will differ from region to region around the globe. Anticipated effects include warming global temperature, rising sea levels, changing precipitation, and expansion of deserts in the subtropics. Warming is expected to be greater over land than over the oceans and greatest in the Arctic, with the continuing retreat of glaciers, permafrost and sea ice. Other likely changes include more frequent extreme weather events including heat waves, droughts, heavy rainfall with floods and heavy snowfall; ocean acidification; and species extinctions due to shifting temperature regimes. Effects significant to humans include the threat to food security from decreasing crop yields and the abandonment of populated areas due to rising sea levels. Because the climate system has a large "inertia" and $CO_2$ will stay in the atmosphere for a long time, many of these effects will not only exist for decades or centuries, but will persist for tens of thousands of years.

Possible societal responses to global warming include mitigation by emissions reduction, adaptation to its effects, building systems resilient to its effects, and possible future climate engineering. Most countries are parties to the United Nations Framework Convention on Climate Change (UNFCCC), whose ultimate objective is to prevent dangerous anthropogenic climate change. Parties to the UNFCCC have agreed that deep cuts in emissions are required and that global warming should be limited to well below 2.0 °C (3.6 °F) relative to pre-industrial levels,with efforts made to limit warming to 1.5 °C (2.7 °F).

Public reactions to global warming and concern about its effects are also increasing. A global 2015 Pew Research Center report showed a median of 54% consider it "a very serious problem". There are significant regional differences, with Americans and Chinese (whose economies are responsible for the greatest annual CO2 emissions) among the least concerned.

## Observed Temperature Changes

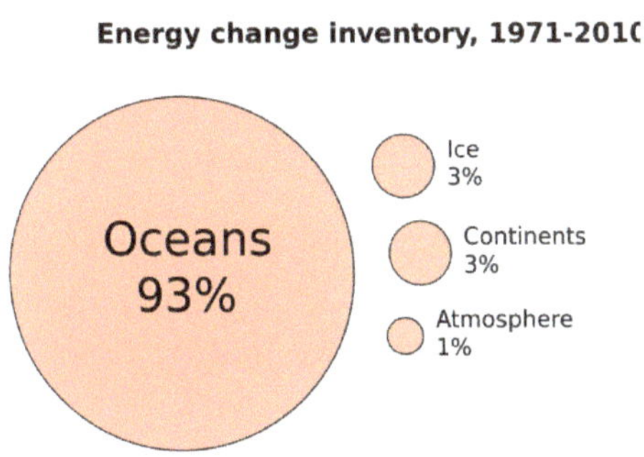


Earth has been in radiative imbalance since at least the 1970s, where less energy leaves the atmosphere than enters it. Most of this extra energy has been absorbed by the oceans. It is very likely that human activities substantially contributed to this increase in ocean heat content.

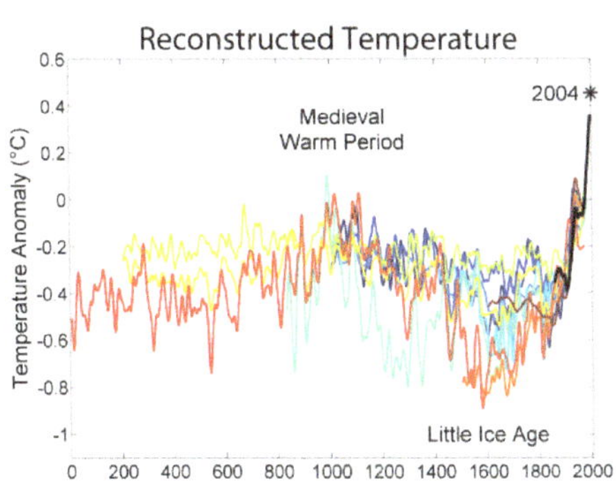


Two millennia of mean surface temperatures according to different reconstructions from climate proxies, each smoothed on a decadal scale, with the instrumental temperature record overlaid in black.

The global average (land and ocean) surface temperature shows a warming of 0.85 [0.65 to 1.06] °C in the period 1880 to 2012, based on multiple independently produced datasets. Earth's average surface temperature rose by 0.74±0.18 °C over the period 1906–2005. The rate of warming almost doubled for the last half of that period (0.13±0.03 °C per decade, versus 0.07±0.02 °C per decade).

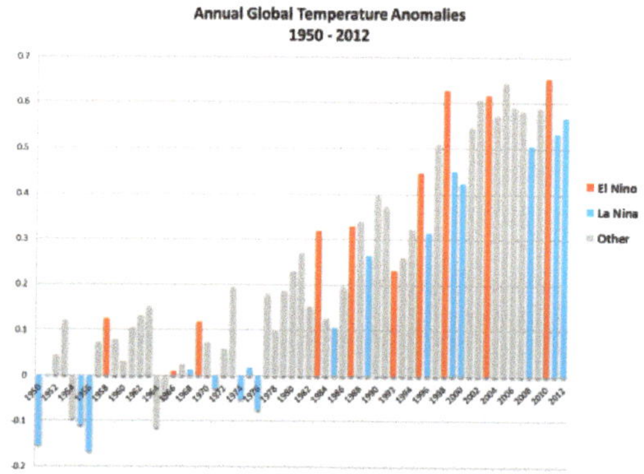


NOAA graph of Global Annual Temperature Anomalies 1950–2012, showing the El Niño Southern Oscillation

The average temperature of the lower troposphere has increased between 0.13 and 0.22 °C (0.23 and 0.40 °F) per decade since 1979, according to satellite temperature measurements. Climate proxies show the temperature to have been relatively stable over the one or two thousand years before 1850, with regionally varying fluctuations such as the Medieval Warm Period and the Little Ice Age.

The warming that is evident in the instrumental temperature record is consistent with a wide range of observations, as documented by many independent scientific groups. Examples include sea level rise, widespread melting of snow and land ice, increased heat content of the oceans, increased humidity, and the earlier timing of spring events, e.g., the flowering of plants. The probability that these changes could have occurred by chance is virtually zero.

## Trends

Temperature changes vary over the globe. Since 1979, land temperatures have increased about twice as fast as ocean temperatures (0.25 °C per decade against 0.13 °C per decade). Ocean temperatures increase more slowly than land temperatures because of the larger effective heat capacity of the oceans and because the ocean loses more heat by evaporation. Since the beginning of industrialisation the temperature difference between the hemispheres has increased due to melting of sea ice and snow in the North. Average arctic temperatures have been increasing at almost twice the rate of the rest of the world in the past 100 years; however arctic temperatures are also highly variable. Although more greenhouse gases are emitted in the Northern than Southern Hemisphere this does not contribute to the difference in warming because the major greenhouse gases persist long enough to mix between hemispheres.

The thermal inertia of the oceans and slow responses of other indirect effects mean that climate can take centuries or longer to adjust to changes in forcing. One climate commitment study concluded that if greenhouse gases were stabilized at year 2000 levels, surface temperatures would still increase by about one-half degree Celsius, and another found that if they were stabilized at 2005 levels surface warming could exceed a whole degree Celsius. Some of this surface warming will be driven by past natural forcings which are still seeking equilibrium in the climate system. One study using a highly simplified climate model indicates these past natural forcings may account for as much as 64% of the committed 2050 surface warming and their influence will fade with time compared to the human contribution.

Global temperature is subject to short-term fluctuations that overlay long-term trends and can temporarily mask them. The relative stability in surface temperature from 2002 to 2009, which has been dubbed the global warming hiatus by the media and some scientists, is consistent with such an episode. 2015 updates to account for differing methods of measuring ocean surface temperature measurements show a positive trend over the recent decade.

## Warmest Years

15 of the top 16 warmest years have occurred since 2000. While record-breaking years can attract considerable public interest, individual years are less significant than the overall trend. So some climatologists have criticized the attention that the popular press gives to "warmest year" statistics; for example, Gavin Schmidt stated "the long-term trends or the expected sequence of records are far more important than whether any single year is a record or not."

2015 was not only the warmest year on record, it broke the record by the largest margin by which the record has been broken. 2015 was the 39th consecutive year with above-average temperatures. Ocean oscillations like El Niño Southern Oscillation (ENSO) can affect global average temperatures, for example, 1998 temperatures were significantly enhanced by strong El Niño conditions. 1998 remained the warmest year until 2005 and 2010 and the temperature of both of these years was enhanced by El Niño periods. The large margin by which 2015 is the warmest year is also attributed to another strong El Niño. However, 2014 was ENSO neutral. According to NOAA and NASA, 2015 had the warmest respective months on record for 10 out of the 12 months. The average temperature around the globe was 1.62°F (0.90°C) or 20% above the twentieth century average. In a first, December 2015 was also the first month to ever reach a temperature 2 degrees Fahrenheit above normal for the planet. As of July 2016, for the 15th consecutive month, the global land and ocean temperature departure from average was the highest since global temperature records began in 1880. This marks the longest such streak in NOAA's 137 years of record keeping. In addition, NOAA reported that July of 2016 was the hottest month on record.

## Initial Causes of Temperature Changes (External Forcings)

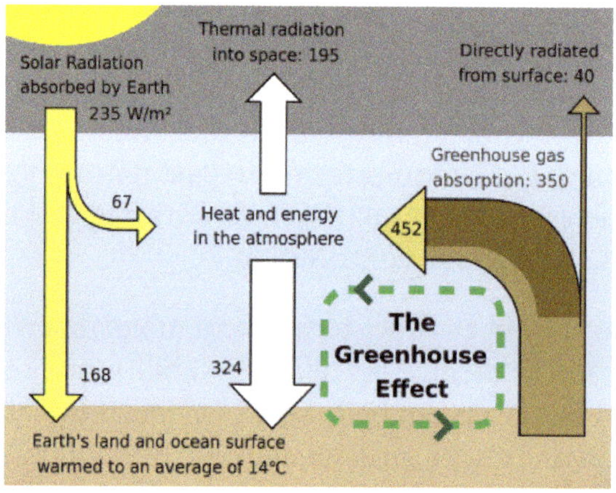


Greenhouse effect schematic showing energy flows between space, the atmosphere, and Earth's surface. Energy exchanges are expressed in watts per square metre (W/m2).

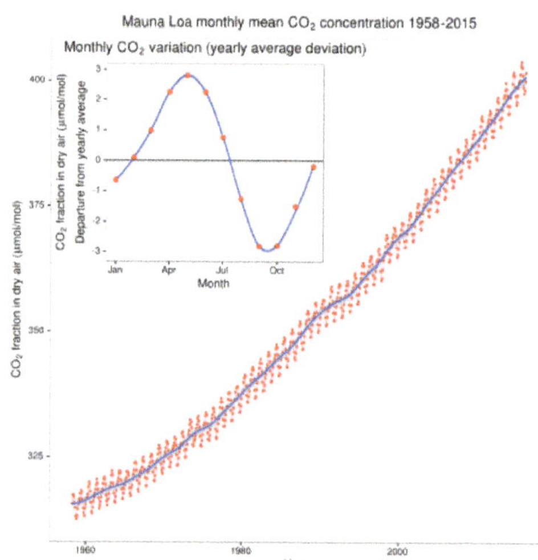


This graph, known as the Keeling Curve, documents the increase of atmospheric carbon dioxide concentrations from 1958–2015. Monthly CO2 measurements display seasonal oscillations in an upward trend; each year's maximum occurs during the Northern Hemisphere's late spring, and declines during its growing season as plants remove some atmospheric CO2.

The climate system can warm or cool in response to changes in *external forcings*. These are "external" to the climate system but not necessarily external to Earth. Examples of external forcings include changes in atmospheric composition (e.g., increased concentrations of greenhouse gases), solar luminosity, volcanic eruptions, and variations in Earth's orbit around the Sun.

## Greenhouse Gases

The greenhouse effect is the process by which absorption and emission of infrared radiation by gases in a planet's atmosphere warm its lower atmosphere and surface. It was proposed by Joseph Fourier in 1824, discovered in 1860 by John Tyndall, was first investigated quantitatively by Svante Arrhenius in 1896, and was developed in the 1930s through 1960s by Guy Stewart Callendar.

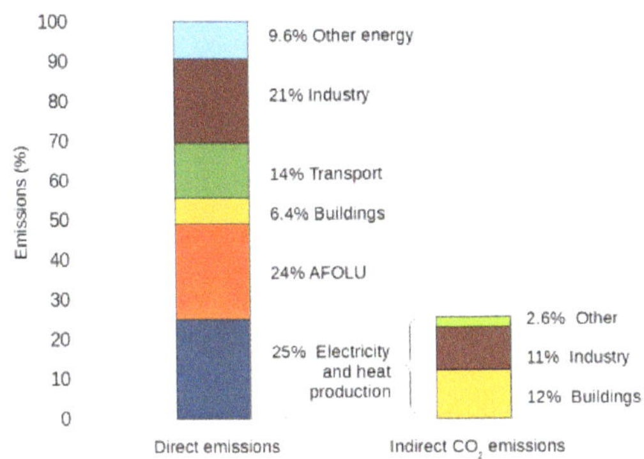


Annual world greenhouse gas emissions, in 2010, by sector.

Percentage share of global cumulative energy-related $CO_2$ emissions between 1751 and 2012 across different regions.

On Earth, naturally occurring amounts of greenhouse gases cause air temperature near the surface to be about 33 °C (59 °F) warmer than it would be in their absence. Without the Earth's atmosphere, the Earth's average temperature would be well below the freezing temperature of water. The major greenhouse gases are water vapour, which causes about 36–70% of the greenhouse effect; carbon dioxide ($CO_2$), which causes 9–26%; methane ($CH_4$), which causes 4–9%; and ozone ($O_3$), which causes 3–7%. Clouds also affect the radiation balance through cloud forcings similar to greenhouse gases.

Human activity since the Industrial Revolution has increased the amount of greenhouse gases in the atmosphere, leading to increased radiative forcing from $CO_2$, methane, tropospheric ozone, CFCs and nitrous oxide. According to work published in 2007, the concentrations of $CO_2$ and methane have increased by 36% and 148% respectively since 1750. These levels are much higher than at any time during the last 800,000 years, the period for which reliable data has been extracted from ice cores. Less direct geological evidence indicates that $CO_2$ values higher than this were last seen about 20 million years ago.

Fossil fuel burning has produced about three-quarters of the increase in $CO_2$ from human activity over the past 20 years. The rest of this increase is caused mostly by changes in land-use, particularly deforestation. Another significant non-fuel source of anthropogenic $CO_2$ emissions is the calcination of limestone for clinker production, a chemical process which releases $CO_2$. Estimates of global $CO_2$ emissions in 2011 from fossil fuel combustion, including cement production and gas flaring, was 34.8 billion tonnes (9.5 ± 0.5 PgC), an increase of 54% above emissions in 1990. Coal burning was responsible for 43% of the total emissions, oil 34%, gas 18%, cement 4.9% and gas flaring 0.7%

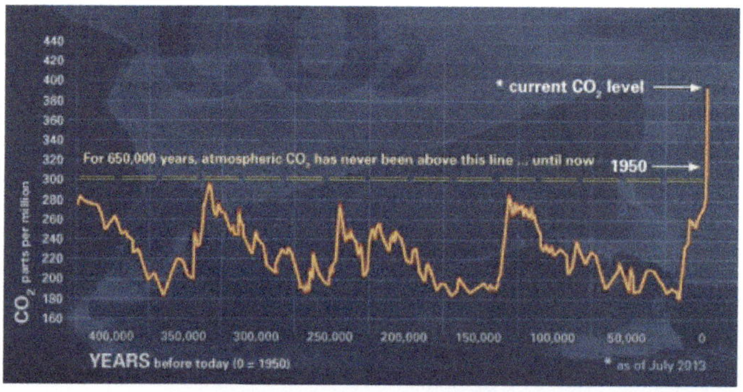

Atmospheric CO2 concentration from 650,000 years ago to near present, using ice core proxy data and direct measurements.

In May 2013, it was reported that readings for $CO_2$ taken at the world's primary benchmark site in Mauna Loa surpassed 400 ppm. According to professor Brian Hoskins, this is likely the first time $CO_2$ levels have been this high for about 4.5 million years. Monthly global $CO_2$ concentrations exceeded 400 ppm in March 2015, probably for the first time in several million years. On 12 November 2015, NASA scientists reported that human-made carbon dioxide continues to increase above levels not seen in hundreds of thousands of years: currently, about half of the carbon dioxide released from the burning of fossil fuels is not absorbed by vegetation and the oceans and remains in the atmosphere.

Over the last three decades of the twentieth century, gross domestic product per capita and population growth were the main drivers of increases in greenhouse gas emissions. $CO_2$ emissions are continuing to rise due to the burning of fossil fuels and land-use change. Emissions can be attributed to different regions. Attributions of emissions due to land-use change are subject to considerable uncertainty.

Emissions scenarios, estimates of changes in future emission levels of greenhouse gases, have been projected that depend upon uncertain economic, sociological, technological, and natural developments. In most scenarios, emissions continue to rise over the century, while in a few, emissions are reduced. Fossil fuel reserves are abundant, and will not limit carbon emissions in the 21st century. Emission scenarios, combined with modelling of the carbon cycle, have been used to produce estimates of how atmospheric concentrations of greenhouse gases might change in the future. Using the six IPCC SRES "marker" scenarios, models suggest that by the year 2100, the atmospheric concentration of $CO_2$ could range between 541 and 970 ppm. This is 90–250% above the concentration in the year 1750.

The popular media and the public often confuse global warming with ozone depletion, i.e., the destruction of stratospheric ozone (e.g., the ozone layer) by chlorofluorocarbons. Although there are a few areas of linkage, the relationship between the two is not strong. Reduced stratospheric ozone has had a slight cooling influence on surface temperatures, while increased tropospheric ozone has had a somewhat larger warming effect.

## Aerosols and Soot

Global dimming, a gradual reduction in the amount of global direct irradiance at the Earth's surface, was observed from 1961 until at least 1990. Solid and liquid particles known as *aerosols*, produced by volcanoes and human-made pollutants, are thought to be the main cause of this dimming. They exert a cooling effect by increasing the reflection of incoming sunlight. The effects of the products of fossil fuel combustion – $CO_2$ and aerosols – have partially offset one another in recent decades, so that net warming has been due to the increase in non-$CO_2$ greenhouse gases such as methane. Radiative forcing due to aerosols is temporally limited due to the processes that remove aerosols from the atmosphere. Removal by clouds and precipitation gives tropospheric aerosols an atmospheric lifetime of only about a week, while stratospheric aerosols can remain for a few years. Carbon dioxide has a lifetime of a century or more, and as such, changes in aerosols will only delay climate changes due to carbon dioxide. Black carbon is second only to carbon dioxide for its contribution to global warming.

Ship tracks can be seen as lines in these clouds over the Atlantic Ocean on the east coast of the United States. Atmospheric particles from these and other sources could have a large effect on climate through the aerosol indirect effect.

In addition to their direct effect by scattering and absorbing solar radiation, aerosols have indirect effects on the Earth's radiation budget. Sulfate aerosols act as cloud condensation nuclei and thus lead to clouds that have more and smaller cloud droplets. These clouds reflect solar radiation more efficiently than clouds with fewer and larger droplets, a phenomenon known as the Twomey effect. This effect also causes droplets to be of more uniform size, which reduces growth of raindrops and makes the cloud more reflective to incoming sunlight, known as the Albrecht effect. Indirect effects are most noticeable in marine stratiform clouds, and have very little radiative effect on convective clouds. Indirect effects of aerosols represent the largest uncertainty in radiative forcing.

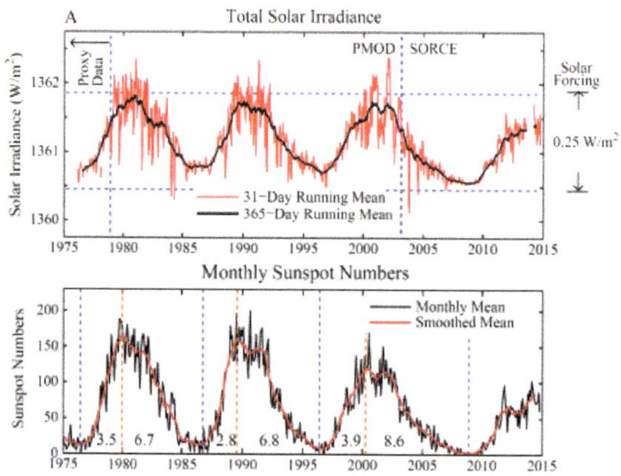


Changes in Total Solar Irradiance (TSI) and monthly sunspot numbers since the mid-1970s.

Soot may either cool or warm Earth's climate system, depending on whether it is airborne or deposited. Atmospheric soot directly absorbs solar radiation, which heats the atmosphere and cools the surface. In isolated areas with high soot production, such as rural India, as much as 50% of surface warming due to greenhouse gases may be masked by atmospheric brown clouds. When deposited, especially on glaciers or on ice in arctic regions, the lower surface albedo can also directly heat the surface. The influences of atmospheric particles, including black carbon, are most pronounced in the tropics and sub-tropics, particularly in Asia, while the effects of greenhouse gases are dominant in the extratropics and southern hemisphere.

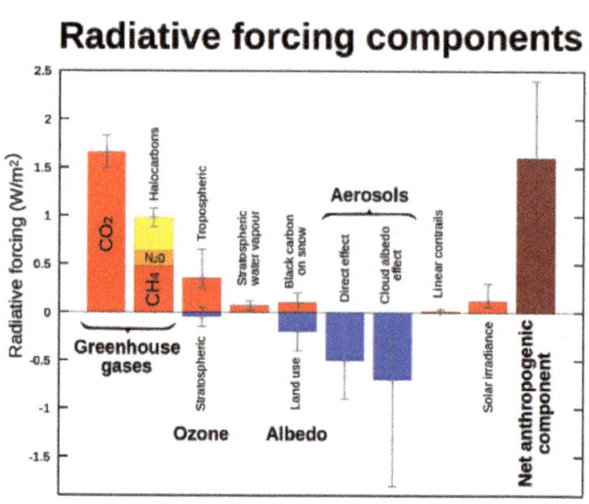


Contribution of natural factors and human activities to radiative forcing of climate change. Radiative forcing values are for the year 2005, relative to the pre-industrial era (1750). The contribution of solar irradiance to radiative forcing is 5% the value of the combined radiative forcing due to increases in the atmospheric concentrations of carbon dioxide, methane and nitrous oxide.

## Solar Activity

Since 1978, solar irradiance has been measured by satellites. These measurements indicate that the Sun's radiative output has not increased since 1978, so the warming during the past 30 years cannot be attributed to an increase in solar energy reaching the Earth.

Climate models have been used to examine the role of the Sun in recent climate change. Models are unable to reproduce the rapid warming observed in recent decades when they only take into account variations in solar output and volcanic activity. Models are, however, able to simulate the observed 20th century changes in temperature when they include all of the most important external forcings, including human influences and natural forcings.

Another line of evidence against solar variations having caused recent climate change comes from looking at how temperatures at different levels in the Earth's atmosphere have changed. Models and observations show that greenhouse warming results in warming of the lower atmosphere (the troposphere) but cooling of the upper atmosphere (the stratosphere). Depletion of the ozone layer by chemical refrigerants has also resulted in a strong cooling effect in the stratosphere. If solar variations were responsible for observed warming, warming of both the troposphere and stratosphere would be expected.

## Variations in Earth's Orbit

The tilt of the Earth's axis and the shape of its orbit around the Sun vary slowly over tens of thousands of years and are a natural source of climate change, by changing the seasonal and latitudinal distribution of solar insolation.

During the last few thousand years, this phenomenon contributed to a slow cooling trend at high latitudes of the Northern Hemisphere during summer, a trend that was reversed by greenhouse-gas-induced warming during the 20th century.

Variations in orbital cycles may initiate a new glacial period in the future, though the timing of this depends on greenhouse gas concentrations as well as the orbital forcing. A new glacial period is not expected within the next 50,000 years if atmospheric $CO_2$ concentration remains above 300 ppm.

## Feedback

The climate system includes a range of *feedbacks*, which alter the response of the system to changes in external forcings. Positive feedbacks increase the response of the climate system to an initial forcing, while negative feedbacks reduce it.

There are a range of feedbacks in the climate system, including water vapour, changes in ice-albedo (snow and ice cover affect how much the Earth's surface absorbs or reflects incoming sunlight), clouds, and changes in the Earth's carbon cycle (e.g., the release of carbon from soil). The main negative feedback is the energy the Earth's surface radiates into space as infrared radiation. According to the Stefan-Boltzmann law, if the absolute temperature (as measured in kelvin) doubles, radiated energy increases by a factor of 16 (2 to the 4th power).

Sea ice, shown here in Nunavut, in northern Canada, reflects more sunshine, while open ocean absorbs more, accelerating melting.

Feedbacks are an important factor in determining the sensitivity of the climate system to increased atmospheric greenhouse gas concentrations. Other factors being equal, a higher *climate sensitivity* means that more warming will occur for a given increase in greenhouse gas forcing. Uncertainty over the effect of feedbacks is a major reason why different climate models project different magnitudes of warming for a given forcing scenario. More research is needed to understand the role of clouds and carbon cycle feedbacks in climate projections.

The IPCC projections previously mentioned span the "likely" range (greater than 66% probability, based on expert judgement) for the selected emissions scenarios. However, the IPCC's projections do not reflect the full range of uncertainty. The lower end of the "likely" range appears to be better constrained than the upper end.

## Climate Models

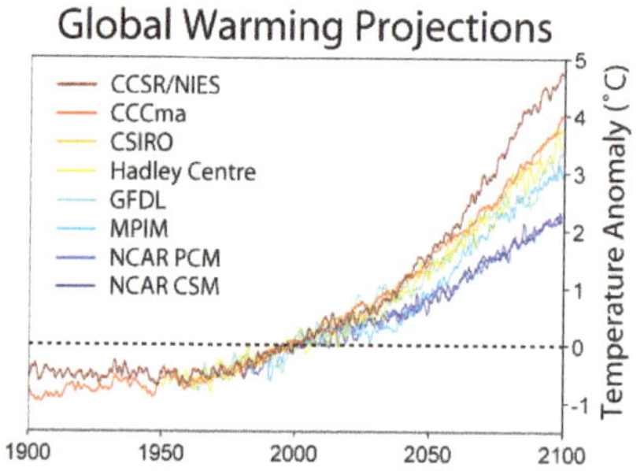


Calculations of global warming prepared in or before 2001 from a range of climate models under the SRES A2 emissions scenario, which assumes no action is taken to reduce emissions and regionally divided economic development.

Projected change in annual mean surface air temperature from the late 20th century to the middle 21st century, based on a medium emissions scenario (SRES A1B). This scenario assumes that no future policies are adopted to limit greenhouse gas emissions. Image credit: NOAA GFDL.

A climate model is a representation of the physical, chemical and biological processes that affect the climate system. Such models are based on scientific disciplines such as fluid dynamics and thermodynamics as well as physical processes such as radiative transfer. The models may be used to predict a range of variables such as local air movement, temperature, clouds, and other atmospheric properties; ocean temperature, salt content, and circulation; ice cover on land and sea; the transfer of heat and moisture from soil and vegetation to the atmosphere; and chemical and biological processes, among others.

Although researchers attempt to include as many processes as possible, simplifications of the actual climate system are inevitable because of the constraints of available computer power and limitations in knowledge of the climate system. Results from models can also vary due to different greenhouse gas inputs and the model's climate sensitivity. For example, the uncertainty in IPCC's 2007 projections is caused by (1) the use of multiple models with differing sensitivity to greenhouse gas concentrations, (2) the use of differing estimates of humanity's future greenhouse gas emissions, (3) any additional emissions from climate feedbacks that were not included in the models IPCC used to prepare its report, i.e., greenhouse gas releases from permafrost.

The models do not assume the climate will warm due to increasing levels of greenhouse gases. Instead the models predict how greenhouse gases will interact with radiative transfer and other physical processes. Warming or cooling is thus a result, not an assumption, of the models.

Clouds and their effects are especially difficult to predict. Improving the models' representation of clouds is therefore an important topic in current research. Another prominent research topic is expanding and improving representations of the carbon cycle.

Models are also used to help investigate the causes of recent climate change by comparing the observed changes to those that the models project from various natural and human causes. Although these models do not unambiguously attribute the warming that occurred from approximately 1910 to 1945 to either natural variation or human effects, they do indicate that the warming since 1970 is dominated by anthropogenic greenhouse gas emissions.

The physical realism of models is tested by examining their ability to simulate contemporary or past climates. Climate models produce a good match to observations of global temperature changes over the last century, but do not simulate all aspects of climate. Not all effects of global warming are accurately predicted by the climate models used by the IPCC. Observed Arctic shrinkage has been faster than that predicted. Precipitation increased proportionally to atmospheric humidity, and hence significantly faster than global climate models predict. Since 1990, sea level has also risen considerably faster than models predicted it would.

## Observed and Expected Environmental Effects

Anthropogenic forcing has likely contributed to some of the observed changes, including sea level rise, changes in climate extremes (such as the number of warm and cold days), declines in Arctic sea ice extent, glacier retreat, and greening of the Sahara.

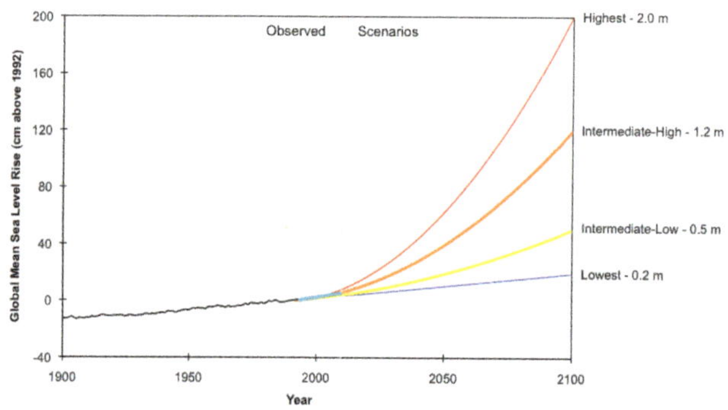


Projections of global mean sea level rise by Parris and others. Probabilities have not been assigned to these projections. Therefore, none of these projections should be interpreted as a "best estimate" of future sea level rise.

During the 21st century, glaciers and snow cover are projected to continue their widespread retreat. Projections of declines in Arctic sea ice vary. Recent projections suggest that Arctic summers could be ice-free (defined as ice extent less than 1 million square km) as early as 2025-2030.

"Detection" is the process of demonstrating that climate has changed in some defined statistical sense, without providing a reason for that change. Detection does not imply attribution of the detected change to a particular cause. "Attribution" of causes of climate change is the process of establishing the most likely causes for the detected change with some defined level of confidence. Detection and attribution may also be applied to observed changes in physical, ecological and social systems.

## Extreme Weather

Changes in regional climate are expected to include greater warming over land, with most warming at high northern latitudes, and least warming over the Southern Ocean and parts of the North Atlantic Ocean.

Future changes in precipitation are expected to follow existing trends, with reduced precipitation over subtropical land areas, and increased precipitation at subpolar latitudes and some equatorial regions. Projections suggest a probable increase in the frequency and severity of some extreme weather events, such as heat waves.

A 2015 study published in *Nature Climate Change*, states:

> "About 18% of the moderate daily precipitation extremes over land are attributable to the observed temperature increase since pre-industrial times, which in turn primarily results from human influence. For 2 °C of warming the fraction of precipitation extremes attributable to human influence rises to about 40%. Likewise, today about 75% of the moderate daily hot extremes over land are attributable to warming. It is the most rare and extreme events for which the largest fraction is anthropogenic, and that contribution increases nonlinearly with further warming."

Data analysis of extreme events from 1960 till 2010 suggests that droughts and heat waves appear simultaneously with increased frequency. Extremely wet or dry events within the monsoon period have increased since 1980.

## Sea Level Rise

The sea level rise since 1993 has been estimated to have been on average 2.6 mm and 2.9 mm per year ± 0.4 mm. Additionally, sea level rise has accelerated from 1995 to 2015. Over the 21st century, the IPCC projects for a high emissions scenario, that global mean sea level could rise by 52–98 cm. The IPCC's projections are conservative, and may underestimate future sea level rise. Other estimates suggest that for the same period, global mean sea level could rise by 0.2 to 2.0 m (0.7–6.6 ft), relative to mean sea level in 1992.

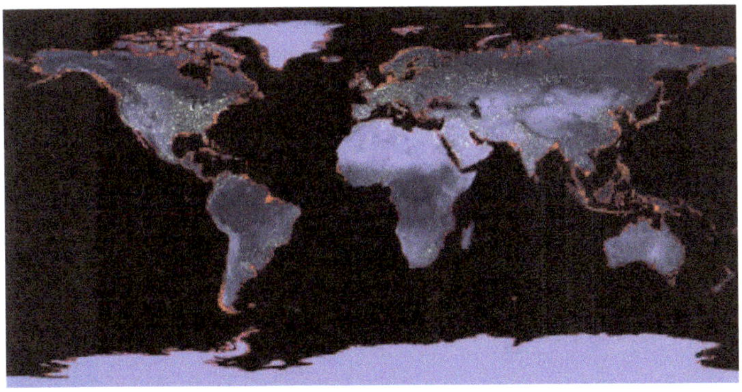

Map of the Earth with a six-metre sea level rise represented in red.

Widespread coastal flooding would be expected if several degrees of warming is sustained for millennia. For example, sustained global warming of more than 2 °C (relative to pre-industrial levels) could lead to eventual sea level rise of around 1 to 4 m due to thermal expansion of sea water and the melting of glaciers and small ice caps. Melting of the Greenland ice sheet could contribute an additional 4 to 7.5 m over many thousands of years. It has been estimated that we are already committed to a sea-level rise of approximately 2.3 metres for each degree of temperature rise within the next 2,000 years.

Warming beyond the 2 °C target would potentially lead to rates of sea-level rise dominated by ice loss from Antarctica. Continued $CO_2$ emissions from fossil sources could cause additional tens of metres of sea level rise, over the next millennia and eventually ultimately eliminate the entire Antarctic ice sheet, causing about 58 metres of sea level rise.

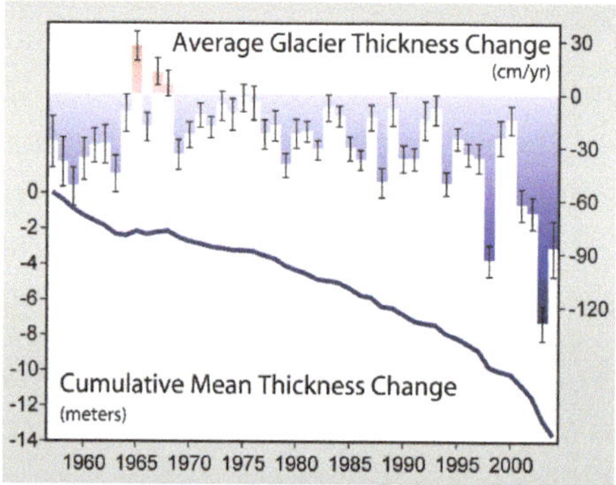


Sparse records indicate that glaciers have been retreating since the early 1800s. In the 1950s measurements began that allow the monitoring of glacial mass balance, reported to the World Glacier Monitoring Service (WGMS) and the National Snow and Ice Data Center (NSIDC).

## Ecological Systems

In terrestrial ecosystems, the earlier timing of spring events, as well as poleward and upward shifts in plant and animal ranges, have been linked with high confidence to recent warming. Future climate change is expected to affect particular ecosystems, including tundra, mangroves, and coral reefs. It is expected that most ecosystems will be affected by higher atmospheric $CO_2$ levels, combined with higher global temperatures. Overall, it is expected that climate change will result in the extinction of many species and reduced diversity of ecosystems.

Increases in atmospheric $CO_2$ concentrations have led to an increase in ocean acidity. Dissolved $CO_2$ increases ocean acidity, measured by lower pH values. Between 1750 and 2000, surface-ocean pH has decreased by ≈0.1, from ≈8.2 to ≈8.1. Surface-ocean pH has probably not been below ≈8.1 during the past 2 million years. Projections suggest that surface-ocean pH could decrease by an additional 0.3–0.4 units by 2100. Future ocean acidification could threaten coral reefs, fisheries, protected species, and other natural resources of value to society.

Ocean deoxygenation is projected to increase hypoxia by 10%, and triple suboxic waters (oxygen concentrations 98% less than the mean surface concentrations), for each 1 °C of upper ocean warming.

## Long-term Effects

On the timescale of centuries to millennia, the magnitude of global warming will be determined primarily by anthropogenic $CO_2$ emissions. This is due to carbon dioxide's very long lifetime in the atmosphere.

Stabilizing the global average temperature would require large reductions in $CO_2$ emissions, as well as reductions in emissions of other greenhouse gases such as methane and nitrous oxide. Emissions of $CO_2$ would need to be reduced by more than 80% relative to their peak level. Even if this were achieved, global average temperatures would remain close to their highest level for many centuries.

Long-term effects also include a response from the Earth's crust, due to ice melting and deglaciation, in a process called post-glacial rebound, when land masses are no longer depressed by the weight of ice. This could lead to landslides and increased seismic and volcanic activities. Tsunamis could be generated by submarine landslides caused by warmer ocean water thawing ocean-floor permafrost or releasing gas hydrates. Some world regions, such as the French Alps, already show signs of an increase in landslide frequency.

## Large-scale and Abrupt Impacts

Climate change could result in global, large-scale changes in natural and social systems. Examples include the possibility for the Atlantic Meridional Overturning Circulation to slow- or shutdown, which in the instance of a shutdown would change weather in Europe and North America considerably, ocean acidification caused by increased atmospheric concentrations of carbon dioxide, and the long-term melting of ice sheets, which contributes to sea level rise.

Some large-scale changes could occur abruptly, i.e., over a short time period, and might also be irreversible. Examples of abrupt climate change are the rapid release of methane and carbon dioxide from permafrost, which would lead to amplified global warming, or the shutdown of thermohaline circulation. Scientific understanding of abrupt climate change is generally poor. The probability of abrupt change for some climate related feedbacks may be low. Factors that may increase the probability of abrupt climate change include higher magnitudes of global warming, warming that occurs more rapidly, and warming that is sustained over longer time periods.

## Observed and Expected Effects on Social Systems

The effects of climate change on human systems, mostly due to warming or shifts in precipitation patterns, or both, have been detected worldwide. Production of wheat and maize globally has been impacted by climate change. While crop production has increased in some mid-latitude regions such as the UK and Northeast China, economic losses due to extreme weather events have increased globally. There has been a shift from cold- to heat-related mortality in some regions as a result of warming. Livelihoods of indigenous peoples of the Arctic have been altered by climate change, and there is emerging evidence of climate change impacts on livelihoods of indigenous peoples in other regions. Regional impacts of climate change are now observable at more locations than before, on all continents and across ocean regions.

The future social impacts of climate change will be uneven. Many risks are expected to increase with higher magnitudes of global warming. All regions are at risk of experiencing negative impacts. Low-latitude, less developed areas face the greatest risk. A study from 2015 concluded that economic growth (gross domestic product) of poorer countries is much more impaired with projected future climate warming, than previously thought.

A meta-analysis of 56 studies concluded in 2014 that each degree of temperature rise will increase violence by up to 20%, which includes fist fights, violent crimes, civil unrest or wars.

Examples of impacts include:

- *Food*: Crop production will probably be negatively affected in low latitude countries, while effects at northern latitudes may be positive or negative. Global warming of around 4.6 °C relative to pre-industrial levels could pose a large risk to global and regional food security.
- *Health*: Generally impacts will be more negative than positive. Impacts include: the effects of extreme weather, leading to injury and loss of life; and indirect effects, such as undernutrition brought on by crop failures.

## Habitat Inundation

In small islands and mega deltas, inundation as a result of sea level rise is expected to threaten vital infrastructure and human settlements. This could lead to issues of homelessness in countries with low-lying areas such as Bangladesh, as well as statelessness for populations in countries such as the Maldives and Tuvalu.

## Economy

Estimates based on the IPCC A1B emission scenario from additional $CO_2$ and $CH_4$ greenhouse gases released from permafrost, estimate associated impact damages by US$43 trillion.

## Infrastructure

Continued permafrost degradation will likely result in unstable infrastructure in Arctic regions, or Alaska before 2100. Thus, impacting roads, pipelines and buildings, as well as water distribution, and cause slope failures.

## Possible Responses to Global Warming

### Mitigation

The graph on the right shows three "pathways" to meet the UNFCCC's 2 °C target, labelled "global technology", "decentralized solutions", and "consumption change".

Each pathway shows how various measures (e.g., improved energy efficiency, increased use of renewable energy) could contribute to emissions reductions. Image credit: PBL Netherlands Environmental Assessment Agency.

Mitigation of climate change are actions to reduce greenhouse gas emissions, or enhance the capacity of carbon sinks to absorb GHGs from the atmosphere. There is a large potential for future reductions in emissions by a combination of activities, including: energy conservation and increased energy efficiency; the use of low-carbon energy technologies, such as renewable energy, nuclear energy, and carbon capture and storage; and enhancing carbon sinks through, for example, reforestation and preventing deforestation. A 2015 report by Citibank concluded that transitioning to a low carbon economy would yield positive return on investments.

Near- and long-term trends in the global energy system are inconsistent with limiting global warming at below 1.5 or 2 °C, relative to pre-industrial levels. Pledges made as part of the Cancún agreements are broadly consistent with having a likely chance (66 to 100% probability) of limiting global warming (in the 21st century) at below 3 °C, relative to pre-industrial levels.

In limiting warming at below 2 °C, more stringent emission reductions in the near-term would allow for less rapid reductions after 2030. Many integrated models are unable to meet the 2 °C target if pessimistic assumptions are made about the availability of mitigation technologies.

## Adaptation

Other policy responses include adaptation to climate change. Adaptation to climate change may be planned, either in reaction to or anticipation of climate change, or spontaneous, i.e., without government intervention. Planned adaptation is already occurring on a limited basis. The barriers, limits, and costs of future adaptation are not fully understood.

A concept related to adaptation is *adaptive capacity*, which is the ability of a system (human, natural or managed) to adjust to climate change (including climate variability and extremes) to moderate potential damages, to take advantage of opportunities, or to cope with consequences. Unmitigated climate change (i.e., future climate change without efforts to limit greenhouse gas emissions) would, in the long term, be likely to exceed the capacity of natural, managed and human systems to adapt.

Environmental organizations and public figures have emphasized changes in the climate and the risks they entail, while promoting adaptation to changes in infrastructural needs and emissions reductions.

## Climate Engineering

Climate engineering (sometimes called *geoengineering* or *climate intervention*) is the deliberate modification of the climate. It has been investigated as a possible response to global warming, e.g. by NASA and the Royal Society. Techniques under research fall generally into the categories solar radiation management and carbon dioxide removal, although various other schemes have been suggested. A study from 2014 investigated the most common climate engineering methods and concluded they are either ineffective or have potentially severe side effects and cannot be stopped without causing rapid climate change.

## Discourse About Global Warming

## Political Discussion

Article 2 of the UN Framework Convention refers explicitly to "stabilization of greenhouse gas concentrations." To stabilize the atmospheric concentration of CO2, emissions worldwide would need to be dramatically reduced from their present level.

Most countries in the world are parties to the United Nations Framework Convention on Climate Change (UNFCCC). The ultimate objective of the Convention is to prevent dangerous human interference of the climate system. As stated in the Convention, this requires that GHG concentrations are stabilized in the atmosphere at a level where ecosystems can adapt naturally to climate change, food production is not threatened, and economic development can proceed in a sustainable fashion. The Framework Convention was agreed in 1992, but since then, global emissions have risen.

During negotiations, the G77 (a lobbying group in the United Nations representing 133 developing nations) pushed for a mandate requiring developed countries to "[take] the lead" in reducing their emissions. This was justified on the basis that: the developed world's emissions had contributed most to the cumulation of GHGs in the atmosphere; per-capita emissions (i.e., emissions per head of population) were still relatively low in developing countries; and the emissions of developing countries would grow to meet their development needs.

This mandate was sustained in the Kyoto Protocol to the Framework Convention, which entered into legal effect in 2005. In ratifying the Kyoto Protocol, most developed countries accepted legally binding commitments to limit their emissions. These first-round commitments expired in 2012. United States President George W. Bush rejected the treaty on the basis that "it exempts 80% of the world, including major population centres such as China and India, from compliance, and would cause serious harm to the US economy."

At the 15th UNFCCC Conference of the Parties, held in 2009 at Copenhagen, several UNFCCC Parties produced the Copenhagen Accord. Parties associated with the Accord (140 countries, as of November 2010) aim to limit the future increase in global mean temperature to below 2 °C. The 16th Conference of the Parties (COP16) was held at Cancún in 2010. It produced an agreement, not a binding treaty, that the Parties should take urgent action to reduce greenhouse gas emissions to meet a goal of limiting global warming to 2 °C above pre-industrial temperatures. It also recognized the need to consider strengthening the goal to a global average rise of 1.5 °C.

## Scientific Discussion

There is continuing discussion through published peer-reviewed scientific papers, which are assessed by scientists working in the relevant fields taking part in the Intergovernmental Panel on Climate Change. The scientific consensus as of 2013 stated in the IPCC Fifth Assessment Report is that it "is extremely likely that human influence has been the dominant cause of the observed warming since the mid-20th century". A 2008 report by the U.S. National Academy of Sciences stated that most scientists by then agreed that observed warming in recent decades was primarily caused by human activities increasing the amount of greenhouse gases in the atmosphere. In 2005 the Royal Society stated that while the overwhelming majority of scientists were in agreement on the main points, some individuals and organizations opposed to the consensus on urgent action needed to reduce greenhouse gas emissions have tried to undermine the science and work of the IPCC. National science academies have called on world leaders for policies to cut global emissions.

In the scientific literature, there is a strong consensus that global surface temperatures have increased in recent decades and that the trend is caused mainly by human-induced emissions of greenhouse gases. No scientific body of national or international standing disagrees with this view.

## Discussion by the Public and in Popular Media

The global warming controversy refers to a variety of disputes, substantially more pronounced in the popular media than in the scientific literature, regarding the nature, causes, and consequences of global warming. The disputed issues include the causes of increased global average air temperature, especially since the mid-20th century, whether this warming trend is unprecedented or within normal climatic variations, whether humankind has contributed significantly to it, and whether the increase is completely or partially an artefact of poor measurements. Additional disputes concern estimates of climate sensitivity, predictions of additional warming, and what the consequences of global warming will be.

From 1990 to 1997, right-wing conservative think tanks in the United States mobilized to challenge the legitimacy of global warming as a social problem. They challenged the scientific evidence, argued that global warming will have benefits, and asserted that proposed solutions would do more harm than good. Some people dispute aspects of climate change science. Organizations such as the libertarian Competitive Enterprise Institute, conservative commentators, and some companies such as ExxonMobil have challenged IPCC climate change scenarios, funded scientists who disagree with the scientific consensus, and provided their own projections of the economic cost of stricter controls. On the other hand, some fossil fuel companies have scaled back their efforts in recent years, or even called for policies to reduce global warming. Global oil companies have begun to acknowledge climate change exists and is caused by human activities and the burning of fossil fuels.

Global warming was the cover story in this 2007 issue of Ms. magazine

## Surveys of Public Opinion

The world public, or at least people in economically advanced regions, became broadly aware of the global warming problem in the late 1980s. Polling groups began to track opinions on the subject, at first mainly in the United States. The longest consistent polling, by Gallup in the US, found relatively small deviations of 10% or so from 1998 to 2015 in opinion on the seriousness of global warming, but with increasing polarization between those concerned and those unconcerned.

The first major worldwide poll, conducted by Gallup in 2008-2009 in 127 countries, found that some 62% of people worldwide said they knew about global warming. In the advanced countries of North America, Europe and Japan, 90% or more knew about it (97% in the U.S., 99% in Japan); in less developed countries, especially in Africa, fewer than a quarter knew about it, although many had noticed local weather changes. Among those who knew about global warming, there was a wide variation between nations in belief that the warming was a result of human activities.

By 2010, with 111 countries surveyed, Gallup determined that there was a substantial decrease since 2007–08 in the number of Americans and Europeans who viewed global warming as a serious threat. In the US, just a little over half the population (53%) now viewed it as a serious concern for either themselves or their families; this was 10 points below the 2008 poll (63%). Latin America had the biggest rise in concern: 73% said global warming is a serious threat to their families. This global poll also found that people are more likely to attribute global warming to human activities than to natural causes, except in the US where nearly half (47%) of the population attributed global warming to natural causes.

A March–May 2013 survey by Pew Research Center for the People & the Press polled 39 countries about global threats. According to 54% of those questioned, global warming featured top of the perceived global threats. In a January 2013 survey, Pew found that 69% of Americans say there is solid evidence that the Earth's average temperature has got warmer over the past few decades, up six points since November 2011 and 12 points since 2009.

A 2010 survey of 14 industrialized countries found that skepticism about the danger of global warming was highest in Australia, Norway, New Zealand and the United States, in that order, correlating positively with per capita emissions of carbon dioxide.

## Etymology

In the 1950s, research suggested increasing temperatures, and a 1952 newspaper reported "climate change". This phrase next appeared in a November 1957 report in *The Hammond Times* which described Roger Revelle's research into the effects of increasing human-caused $CO_2$ emissions on the greenhouse effect, "a large scale global warming, with radical climate changes may result". Both phrases were only used occasionally until 1975, when Wallace Smith Broecker published a scientific paper on the topic; "Climatic Change: Are We on the Brink of a Pronounced Global Warming?" The phrase began to come into common use, and in 1976 Mikhail Budyko's statement that "a global warming up has started" was widely reported. Other studies, such as a 1971 MIT report, referred to the human impact as "inadvertent climate modification", but an influential 1979 National Academy of Sciences study headed by Jule Charney followed Broecker in using *global warming* for rising surface temperatures, while describing the wider effects of increased $CO_2$ as *climate change*.

In 1986 and November 1987, NASA climate scientist James Hansen gave testimony to Congress on global warming. There were increasing heatwaves and drought problems in the summer of 1988, and when Hansen testified in the Senate on 23 June he sparked worldwide interest. He said: "global warming has reached a level such that we can ascribe with a high degree of confidence a cause and effect relationship between the greenhouse effect and the observed warming." Public attention increased over the summer, and *global warming* became the dominant popular term, commonly used both by the press and in public discourse.

In a 2008 NASA article on usage, Erik M. Conway defined *Global warming* as "the increase in Earth's average surface temperature due to rising levels of greenhouse gases", while *Climate change* was "a long-term change in the Earth's climate, or of a region on Earth." As effects such as changing patterns of rainfall and rising sea levels would probably have more impact than temperatures alone, he considered *global climate change* a more scientifically accurate term, and like the Intergovernmental Panel on Climate Change, the NASA website would emphasize this wider context.

# Environmental Resource Management

Environmental resource management is the management of the interaction and impact of human societies on the environment. It is not, as the phrase might suggest, the management of the *environment* itself. Environmental resources management aims to ensure that ecosystem services are protected and maintained for future human generations, and also maintain ecosystem integrity through considering ethical, economic, and scientific (ecological) variables. Environmental resource management tries to identify factors affected by conflicts that rise between meeting needs and protecting resources. It is thus linked to environmental protection, sustainability and integrated landscape management.

The shrinking Aral Sea, an example of poor water resource management diverted for irrigation.

## Significance

Environmental resource management is an issue of increasing concern, as reflected in its prevalence in seminal texts influencing global socio-political frameworks such as the Brundtland Commission's Our Common Future, which highlighted the integrated nature of environment and international development and the Worldwatch Institute's annual State of the World reports.

The environment determines the nature of people, animals, plants, and places around the globe, affecting behaviour, religion, culture and economic practices.

## Scope

Environmental resource management can be viewed from a variety of perspectives. It involves the management of all components of the biophysical environment, both living (biotic) and non-living (abiotic), and the relationships amongst all living species and their habitats. The environment also involves the relationships of the human environment, such as the social, cultural and economic environment, with the biophysical environment. The essential aspects of environmental resource management are ethical, economical, social, and technological. These underlie principles and help make decisions.

Improved agricultural practices such as these terraces in northwest Iowa can serve to preserve soil and improve water quality

The concept of environmental determinism, probabilism and possibilism are significant in the concept of environmental resource management.

It should be noted that environmental resource management covers many areas in science, including geography, biology, physics, chemistry, sociology, psychology, and physiology.

## Aspects

### Ethical

Environmental resource management strategies are intrinsically driven by conceptions of human-nature relationships. Ethical aspects involve the cultural and social issues relating to the environment, and dealing with changes to it. "All human activities take place in the context of certain types of relationships between society and the bio-physical world (the rest of nature)," and so, there is a great significance in understanding the ethical values of different groups around the world. Broadly speaking, two schools of thought exist in environmental ethics: Anthropocentrism and Ecocentrism, each influencing a broad spectrum of environmental resource management styles along a continuum. These styles perceive "...different evidence, imperatives, and problems, and prescribe different solutions, strategies, technologies, roles for economic sectors, culture, governments, and ethics, etc."

## Anthropocentrism

Anthropocentrism, "...an inclination to evaluate reality exclusively in terms of human values," is an ethic reflected in the major interpretations of Western religions and the dominant economic paradigms of the industrialised world. Anthropocentrism looks at nature as existing solely for the benefit of humans, and as a commodity to use for the good of humanity and to improve human quality of life. Anthropocentric environmental resource management is therefore not the conservation of the environment solely for the environment's sake, but rather the conservation of the environment, and ecosystem structure, for humans' sake.

## Ecocentrism

Ecocentrists believe in the intrinsic value of nature while maintaining that human beings must use and even exploit nature to survive and live. It is this fine ethical line that ecocentrists navigate between fair use and abuse. At an extreme of the ethical scale, ecocentrism includes philosophies such as ecofeminism and deep ecology, which evolved as a reaction to dominant anthropocentric paradigms. "In its current form, it is an attempt to synthesize many old and some new philosophical attitudes about the relationship between nature and human activity, with particular emphasis on ethical, social, and spiritual aspects that have been downplayed in the dominant economic worldview."

## Economics

The economy functions within, and is dependent upon goods and services provided by natural ecosystems. The role of the environment is recognized in both classical economics and neoclassical economics theories, yet the environment was a lower priority in economic policies from 1950 to 1980 due to emphasis from policy makers on economic growth. With the prevalence of environmental problems, many economists embraced the notion that, "If environmental sustainability must coexist for economic sustainability, then the overall system must [permit] identification of an equilibrium between the environment and the economy." As such, economic policy makers began to incorporate the functions of the natural environment—or natural capital – particularly as a sink for wastes and for the provision of raw materials and amenities.

Debate continues among economists as to how to account for natural capital, specifically whether resources can be replaced through knowledge and technology, or whether the environment is a closed system that cannot be replenished and is finite. Economic models influence environmental resource management, in that management policies reflect beliefs about natural capital scarcity. For someone who believes natural capital is infinite and easily substituted, environmental management is irrelevant to the economy. For example, economic paradigms based on neoclassical models of closed economic systems are primarily concerned with resource scarcity, and thus prescribe legalizing the environment as an economic externality for an environmental resource management strategy. This approach has often been termed 'Command-and-control'. Colby has identified trends in the development of economic paradigms, among them, a shift towards more ecological economics since the 1990s.

A water harvesting system collects rainwater from the Rock of Gibraltar into pipes that lead to tanks excavated inside the rock.

## Ecology

"The pairing of significant uncertainty about the behaviour and response of ecological systems with urgent calls for near-term action constitutes a difficult reality, and a common lament" for many environmental resource managers. Scientific analysis of the environment deals with several dimensions of ecological uncertainty. These include: *structural uncertainty* resulting from the misidentification, or lack of information pertaining to the relationships between ecological variables; *parameter uncertainty* referring to "uncertainty associated with parameter values that are not known precisely but can be assessed and reported in terms of the likelihood...of experiencing a defined range of outcomes"; and *stochastic uncertainty* stemming from chance or unrelated factors. Adaptive management is considered a useful framework for dealing with situations of high levels of uncertainty though it is not without its detractors.

A diagram showing the juvenile fish bypass system, which allows young salmon and steelhead to safely pass the Rocky Reach Hydro Project in Washington

Fencing separates big game from vehicles along the Quebec Autoroute 73 in Canada.

A common scientific concept and impetus behind environmental resource management is carrying capacity. Simply put, carrying capacity refers to the maximum number of organisms a particular resource can sustain. The concept of carrying capacity, whilst understood by many cultures over history, has its roots in Malthusian theory. An example is visible in the EU Water Framework Directive. However, "it is argued that Western scientific knowledge ... is often insufficient to deal with the full complexity of the interplay of variables in environmental resource management. These concerns have been recently addressed by a shift in environmental resource management approaches to incorporate different knowledge systems including traditional knowledge, reflected in approaches such as adaptive co-management community-based natural resource management and transitions management. among others.

## Sustainability

Sustainability in environmental resource management involves managing economic, social, and ecological systems both within and outside an organizational entity so it can sustain itself and the system it exists in. In context, sustainability implies that rather than competing for endless growth on a finite planet, development improves quality of life without necessarily consuming more resources. Sustainably managing environmental resources requires organizational change that instills sustainability values that portrays these values outwardly from all levels and reinforces them to surrounding stakeholders. The end result should be a symbiotic relationship between the sustaining organization, community, and environment.

Many drivers compel environmental resource management to take sustainability issues into account. Today's economic paradigms do not protect the natural environment, yet they deepen human dependency on biodiversity and ecosystem services. Ecologically, massive environmental degradation and climate change threaten the stability of ecological systems that humanity depends on. Socially, an increasing gap between rich and poor and the global North-South divide denies many access to basic human needs, rights, and education, leading to further environmental destruction. The planet's unstable condition is caused by many anthropogenic sources. As an exceptionally powerful contributing factor to social and environmental change, the modern organisation has the potential to apply environmental resource management with sustainability principals to achieve highly effective outcomes. To achieve sustainable development with environmental resource management an organisation should work within sustainability principles, including social and environmental accountability, long-term planning; a strong, shared vision; a holistic focus; devolved and consensus decision making; broad stakeholder engagement and justice; transparency measures; trust; and flexibility.

## Current Paradigm Shifts

To adjust to today's environment of quick social and ecological changes, some organizations have begun to experiment with new tools and concepts. Those that are more traditional and stick to hierarchical decision making have difficulty dealing with the demand for lateral decision making that supports effective participation. Whether it be a matter of ethics or just strategic advantage organizations are internalizing sustainability principles. Some of the world's largest and most profitable corporations are shifting to sustainable environmental resource management: Ford, Toyota, BMW, Honda, Shell, Du Pont, Swiss Re, Hewlett-Packard, and Unilever, among others. An extensive study by the Boston Consulting Group reaching 1,560 business leaders from diverse regions, job positions, expertise in sustainability, industries, and sizes of organizations, revealed the many benefits of sustainable practice as well as its viability.

It is important to note that though sustainability of environmental resource management has improved, corporate sustainability, for one, has yet to reach the majority of global companies operating in the markets. The three major barriers to preventing organizations to shift towards sustainable practice with environmental resource management are not understanding what sustainability is; having difficulty modeling an economically viable case for the switch; and having a flawed execution plan, or a lack thereof. Therefore, the most important part of shifting an organization to adopt sustainability in environmental resource management would be to create a shared vision and understanding of what sustainability is for that particular organization, and to clarify the business case.

## Stakeholders

### Public Sector

The public sector comprises the general government sector plus all public corporations including the central bank. In environmental resource management the public sector is responsible for administering natural resource management and implementing environmental protection legislation. The traditional role of the public sector in environmental resource management is to provide professional judgement through skilled technicians on behalf of the public. With the increase of intractable environmental problems, the public sector has been led to examine alternative paradigms for managing environmental resources. This has resulted in the public sector working collaboratively with other sectors (including other governments, private and civil) to encourage sustainable natural resource management behaviours.

A conservation project in North Carolina involving the search for bog turtles was conducted by United States Fish and Wildlife Service and the North Carolina Wildlife Resources Commission and its volunteers

## Private Sector

The private sector comprises private corporations and non-profit institutions serving households. The private sector's traditional role in environmental resource management is that of the recovery of natural resources. Such private sector recovery groups include mining (minerals and petroleum), forestry and fishery organisations. Environmental resource management undertaken by the private sectors varies dependent upon the resource type, that being renewable or non-renewable and private and common resources. Environmental managers from the private sector also need skills to manage collaboration within a dynamic social and political environment.

## Civil Society

Civil society comprises associations in which societies voluntarily organise themselves into and which represent a wide range of interests and ties. These can include community-based organisations, indigenous peoples' organisations and non-government organisations (NGO). Functioning through strong public pressure, civil society can exercise their legal rights against the implementation of resource management plans, particularly land management plans. The aim of civil society in environmental resource management is to be included in the decision-making process by means of public participation. Public participation can be an effective strategy to invoke a sense of social responsibility of natural resources.

## Tools

As with all management functions, effective management tools, standards and systems are required. An environmental management standard or system or protocol attempts to reduce environmental impact as measured by some objective criteria. The ISO 14001 standard is the most widely used standard for environmental risk management and is closely aligned to the European Eco-Management and Audit Scheme (EMAS). As a common auditing standard, the ISO 19011 standard explains how to combine this with quality management.

Other environmental management systems (EMS) tend to be based on the ISO 14001 standard and many extend it in various ways:

- The Green Dragon Environmental Management Standard is a five-level EMS designed for smaller organisations for whom ISO 14001 may be too onerous and for larger organisations who wish to implement ISO 14001 in a more manageable step-by-step approach,
- BS 8555 is a phased standard that can help smaller companies move to ISO 14001 in six manageable steps,
- The Natural Step focuses on basic sustainability criteria and helps focus engineering on reducing use of materials or energy use that is unsustainable in the long term,
- Natural Capitalism advises using accounting reform and a general biomimicry and industrial ecology approach to do the same thing,
- US Environmental Protection Agency has many further terms and standards that it defines as appropriate to large-scale EMS,
- The UN and World Bank has encouraged adopting a "natural capital" measurement and management framework,
- The European Union Eco-Management and Audit Scheme (EMAS).

Other strategies exist that rely on making simple distinctions rather than building top-down management "systems" using performance audits and full cost accounting. For instance, Ecological Intelligent Design divides products into consumables, service products or durables and unsaleables – toxic products that no one should buy, or in many cases, do not realize they are buying. By eliminating the unsaleables from the comprehensive outcome of any purchase, better environmental resource management is achieved without *systems*.

Recent successful cases have put forward the notion of *integrated management*. It shares a wider approach and stresses out the importance of interdisciplinary assessment. It is an interesting notion that might not be adaptable to all cases.

## Waste Management

Waste management is all the activities and actions required to manage waste from its inception to its final disposal. This includes amongst other things, collection, transport, treatment and disposal of waste together with monitoring and regulation. It also encompasses the legal and regulatory framework that relates to waste management encompassing guidance on recycling etc.

Waste management in Kathmandu, Nepal

The term usually relates to all kinds of waste, whether generated during the extraction of raw materials, the processing of raw materials into intermediate and final products, the consumption of final products, or other human activities, including municipal (residential, institutional, commercial), agricultural, and social (health care, household hazardous wastes, sewage sludge). Waste management is intended to reduce adverse effects of waste on health, the environment or aesthetics.

Waste management in Stockholm, Sweden

Waste management practices are not uniform among countries (developed and developing nations); regions (urban and rural area), and sectors (residential and industrial).

## Central Principles of Waste Management

There are a number of concepts about waste management which vary in their usage between countries or regions. Some of the most general, widely used concepts include:

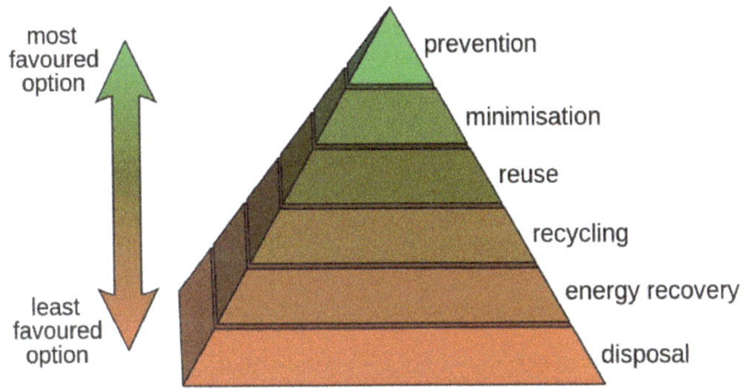


Diagram of the waste hierarchy

## Waste Hierarchy

The waste hierarchy refers to the "3 Rs" reduce, reuse and recycle, which classify waste management strategies according to their desirability in terms of waste minimisation. The waste hierarchy remains the cornerstone of most waste minimisation strategies. The aim of the waste hierarchy is to extract the maximum practical benefits from products and to generate the minimum amount of waste.The waste hierarchy is represented as a pyramid because the basic premise is for policy to take action first and prevent the generation of waste. The next step or preferred action is to reduce the generation of waste i.e. by re-use. The next is recycling which would include composting. Following this step is material recovery and waste-to-energy. Energy can be recovered from processes i.e. landfill and combustion, at this level of the hierarchy. The final action is disposal, in landfills or through incineration without energy recovery. This last step is the final resort for waste which has not been prevented, diverted or recovered. The waste hierarchy represents the progression of a product or material through the sequential stages of the pyramid of waste management. The hierarchy represents the latter parts of the life-cycle for each product.

## Life-cycle of a Product

The life-cycle begins with design, then proceeds through manufacture, distribution, use and then follows through the waste hierarchy's stages of reuse, recovery, recycling and disposal. Each of the above stages of the life-cycle offers opportunities for policy intervention, to rethink the need for the product, to redesign to minimize waste potential, to extend its use. The key behind the life-cycle of a product is to optimize the use of the world's limited resources by avoiding the unnecessary generation of waste.

## Resource Efficiency

Resource efficiency reflects the understanding that current, global, economic growth and development can not be sustained with the current production and consumption patterns. Globally, we are extracting more resources to produce goods than the planet can replenish. Resource efficiency is the reduction of the environmental impact from the production and consumption of these goods, from final raw material extraction to last use and disposal. This process of resource efficiency can address sustainability.

## Polluter Pays Principle

The Polluter pays principle is a principle where the polluting party pays for the impact caused to the environment. With respect to waste management, this generally refers to the requirement for a waste generator to pay for appropriate disposal of the unrecoverable material.

## History

Throughout most of history, the amount of waste generated by humans was insignificant due to low population density and low societal levels of the exploitation of natural resources. Common waste produced during pre-modern times was mainly ashes and human biodegradable waste, and these were released back into the ground locally, with minimum environmental impact. Tools made out of wood or metal were generally reused or passed down through the generations.

However, some civilizations do seem to have been more profligate in their waste output than others. In particular, the Maya of Central America had a fixed monthly ritual, in which the people of the village would gather together and burn their rubbish in large dumps.

## Modern Era

Following the onset of industrialisation and the sustained urban growth of large population centres in England, the buildup of waste in the cities caused a rapid deterioration in levels of sanitation and the general quality of urban life. The streets became choked with filth due to the lack of waste clearance regulations. Calls for the establishment of a municipal authority with waste removal powers occurred as early as 1751, when Corbyn Morris in London proposed that "...as the preservation of the health of the people is of great importance, it is proposed that the cleaning of this city, should be put under one uniform public management, and all the filth be...conveyed by the Thames to proper distance in the country".

However, it was not until the mid-19th century, spurred by increasingly devastating cholera outbreaks and the emergence of a public health debate that the first legislation on the issue emerged. Highly influential in this new focus was the report *The Sanitary Condition of the Labouring Population* in 1842 of the social reformer, Edwin Chadwick, in which he argued for the importance of adequate waste removal and management facilities to improve the health and wellbeing of the city's population.

Sir Edwin Chadwick's 1842 report The Sanitary Condition of the Labouring Population was influential in securing the passage of the first legislation aimed at waste clearance and disposal.

In the UK, the Nuisance Removal and Disease Prevention Act of 1846 began what was to be a steadily evolving process of the provision of regulated waste management in London. The Metropolitan Board of Works was the first city-wide authority that centralized sanitation regulation for the rapidly expanding city and the Public Health Act 1875 made it compulsory for every household to deposit their weekly waste in "moveable receptacles: for disposal—the first concept for a dust-bin.

The dramatic increase in waste for disposal led to the creation of the first incineration plants, or, as they were then called, "destructors". In 1874, the first incinerator was built in Nottingham by Manlove, Alliott & Co. Ltd. to the design of Albert Fryer. However, these were met with opposition on account of the large amounts of ash they produced and which wafted over the neighbouring areas.

Similar municipal systems of waste disposal sprung up at the turn of the 20th century in other large cities of Europe and North America. In 1895, New York City became the first U.S. city with public-sector garbage management.

Early garbage removal trucks were simply open bodied dump trucks pulled by a team of horses. They became motorized in the early part of the 20th century and the first close body trucks to eliminate odours with a dumping lever mechanism were introduced in the 1920s in Britain. These were soon equipped with 'hopper mechanisms' where the scooper was loaded at floor level and then hoisted mechanically to deposit the waste in the truck. The Garwood Load Packer was the first truck in 1938, to incorporate a hydraulic compactor.

Manlove, Alliott & Co. Ltd. 1894 destructor furnace. The use of incinerators for waste disposal became popular in the late 19th century.

## Waste Handling and Transport

Waste collection methods vary widely among different countries and regions. Domestic waste collection services are often provided by local government authorities, or by private companies for industrial and commercial waste. Some areas, especially those in less developed countries, do not have formal waste-collection systems.

Molded plastic, wheeled waste bin in Berkshire, England

## Waste Handling Practices

Curbside collection is the most common method of disposal in most European countries, Canada, New Zealand and many other parts of the developed world in which waste is collected at regular intervals by specialised trucks. This is often associated with curb-side waste segregation. In rural areas waste may need to be taken to a transfer station. Waste collected is then transported to an appropriate disposal facility. In some areas, vacuum collection is used in which waste is transported from the home or commercial premises by vacuum along small bore tubes. Systems are in use in Europe and North America.

Pyrolysis is used to dispose of some wastes including tires, a process that can produce recovered fuels, steel and heat. In some cases tires can provide the feedstock for cement manufacture. Such systems are used in USA, California, Australia, Greece, Mexico, the United Kingdom and in Israel. The RESEM pyrolysis plant that has been operational at Texas USA since December 2011, and processes up to 60 tons per day. In some jurisdictions unsegregated waste is collected at the curb-side or from waste transfer stations and then sorted into recyclables and unusable waste. Such systems are capable of sorting large volumes of solid waste, salvaging recyclables, and turning the rest into bio-gas and soil conditioner. In San Francisco, the local government established its Mandatory Recycling and Composting Ordinance in support of its goal of zero waste by 2020, requiring everyone in the city to keep recyclables and compostables out of the landfill. The three streams are collected with the curbside "Fantastic 3" bin system - blue for recyclables, green for compostables, and black for landfill-bound materials - provided to residents and businesses and serviced by San Francisco's sole refuse hauler, Recology. The City's "Pay-As-You-Throw" system charges customers by the volume of landfill-bound materials, which provides a financial incentive to separate recyclables and compostables from other discards. The City's Department of the Environment's Zero Waste Program has led the City to achieve 80% diversion, the highest diversion rate in North America.

## Financial Models

In most developed countries, domestic waste disposal is funded from a national or local tax which may be related to income, or notional house value. Commercial and industrial waste disposal is typically charged for as a commercial service, often as an integrated charge which includes disposal costs. This practice may encourage disposal contractors to opt for the cheapest disposal option such as landfill rather than the environmentally best solution such as re-use and recycling. In some areas such as Taipei, the city government charges its households and industries for the volume of rubbish they produce. Waste will only be collected by the city council if waste is disposed in government issued rubbish bags. This policy has successfully reduced the amount of waste the city produces and increased the recycling rate.

## Disposal Solutions

### Landfill

Spittelau incineration plant in Vienna

A landfill compaction vehicle in action.

## Incineration

Incineration is a disposal method in which solid organic wastes are subjected to combustion so as to convert them into residue and gaseous products. This method is useful for disposal of residue of both solid waste management and solid residue from waste water management. This process reduces the volumes of solid waste to 20 to 30 percent of the original volume. Incineration and other high temperature waste treatment systems are sometimes described as "thermal treatment". Incinerators convert waste materials into heat, gas, steam, and ash.

Incineration is carried out both on a small scale by individuals and on a large scale by industry. It is used to dispose of solid, liquid and gaseous waste. It is recognized as a practical method of disposing of certain hazardous waste materials (such as biological medical waste). Incineration is a controversial method of waste disposal, due to issues such as emission of gaseous pollutants.

Incineration is common in countries such as Japan where land is more scarce, as these facilities generally do not require as much area as landfills. Waste-to-energy (WtE) or energy-from-waste (EfW) are broad terms for facilities that burn waste in a furnace or boiler to generate heat, steam or electricity. Combustion in an incinerator is not always perfect and there have been concerns about pollutants in gaseous emissions from incinerator stacks. Particular concern has focused on some very persistent organic compounds such as dioxins, furans, and PAHs, which may be created and which may have serious environmental consequences.

## Recycling

Recycling is a resource recovery practice that refers to the collection and reuse of waste materials such as empty beverage containers. The materials from which the items are made can be reprocessed into new products. Material for recycling may be collected separately from general waste using dedicated bins and collection vehicles, a procedure called kerbside collection. In some communities, the owner of the waste is required to separate the materials into different bins (e.g. for paper, plastics, metals) prior to its collection. In other communities, all recyclable materials are placed in a single bin for collection, and the sorting is handled later at a central facility. The latter method is known as "single-stream recycling."

Waste not the Waste. Sign in Tamil Nadu, India

The most common consumer products recycled include aluminium such as beverages cans, copper such as wire, steel from food and aerosol cans, old steel furnishings or equipment, rubber tyres, polyethylene and PET bottles, glass bottles and jars, paperboard cartons, newspapers, magazines and light paper, and corrugated fiberboard boxes.

Steel crushed and baled for recycling

PVC, LDPE, PP, and PS are also recyclable. These items are usually composed of a single type of material, making them relatively easy to recycle into new products. The recycling of complex products (such as computers and electronic equipment) is more difficult, due to the additional dismantling and separation required.

The type of material accepted for recycling varies by city and country. Each city and country has different recycling programs in place that can handle the various types of recyclable materials. However, certain variation in acceptance is reflected in the resale value of the material once it is reprocessed.

## Re-Use

## Biological Reprocessing

An active compost heap.

Recoverable materials that are organic in nature, such as plant material, food scraps, and paper products, can be recovered through composting and digestion processes to decompose the organic matter. The resulting organic material is then recycled as mulch or compost for agricultural or landscaping purposes. In addition, waste gas from the process (such as methane) can be captured and used for generating electricity and heat (CHP/cogeneration) maximising efficiencies. The intention of biological processing in waste management is to control and accelerate the natural process of decomposition of organic matter.

## Energy Recovery

Energy recovery from waste is the conversion of non-recyclable waste materials into usable heat, electricity, or fuel through a variety of processes, including combustion, gasification, pyrolyzation, anaerobic digestion, and landfill gas recovery. This process is often called waste-to-energy. Energy recovery from waste is part of the non-hazardous waste management hierarchy. Using energy recovery to convert non-recyclable waste materials into electricity and heat, generates a renewable energy source and can reduce carbon emissions by offsetting the need for energy from fossil sources as well as reduce methane generation from landfills. Globally, waste-to-energy accounts for 16% of waste management.

The energy content of waste products can be harnessed directly by using them as a direct combustion fuel, or indirectly by processing them into another type of fuel. Thermal treatment ranges from using waste as a fuel source for cooking or heating and the use of the gas fuel, to fuel for boilers to generate steam and electricity in a turbine. Pyrolysis and gasification are two related forms of thermal treatment where waste materials are heated to high temperatures with limited oxygen availability. The process usually occurs in a sealed vessel under high pressure. Pyrolysis of solid waste converts the material into solid, liquid and gas products. The liquid and gas can be burnt to produce energy or refined into other chemical products (chemical refinery). The solid residue (char) can be further refined into products such as activated carbon. Gasification and advanced Plasma arc gasification are used to convert organic materials directly into a synthetic gas (syngas) composed of carbon monoxide and hydrogen. The gas is then burnt to produce electricity and steam. An alternative to pyrolysis is high temperature and pressure supercritical water decomposition (hydrothermal monophasic oxidation).

## Pyrolysis

Pyrolysis is a process of thermo-chemical decomposition of organic materials by heat in the absence of oxygen which produces various hydrocarbon gases. During pyrolysis, the molecules of object are subjected to very high temperatures leading to very high vibrations. Therefore, every molecule in the object is stretched and shaken to an extent that molecules starts breaking down. The rate of pyrolysis increases with temperature. In industrial applications, temperatures are above 430 °C (800 °F). Fast pyrolysis produces liquid fuel for feedstocks like wood. Slow pyrolysis produces gases and solid charcoal. Pyrolysis hold promise for conversion of waste biomass into useful liquid fuel. Pyrolysis of waste plastics can produce millions of litres of fuel. Solid products of this process contain metals, glass, sand and pyrolysis coke which cannot be converted to gas in the process.

## Resource Recovery

Resource recovery is the systematic diversion of waste, which was intended for disposal, for a specific next use. It is the processing of recyclables to extract or recover materials and resources, or convert to energy. These activities are performed at a resource recovery facility. Resource recovery is not only environmentally important, but it is also cost effective. It decreases the amount of waste for disposal, saves space in landfills, and conserves natural resources.

Resource recovery (as opposed to waste management) uses LCA (life cycle analysis) attempts to offer alternatives to waste management. For mixed MSW (Municipal Solid Waste) a number of broad studies have indicated that administration, source separation and collection followed by reuse and recycling of the non-organic fraction and energy and compost/fertilizer production of the organic material via anaerobic digestion to be the favoured path.

As an example of how resource recycling can be beneficial, many of the items thrown away contain precious metals which can be recycled to create a profit, such as the components in circuit boards. Other industries can also benefit from resource recycling with the wood chippings in pallets and other packaging materials being passed onto sectors such as the horticultural profession. In this instance, workers can use the recycled chips to create paths, walkways, or arena surfaces.

## Sustainability

The management of waste is a key component in a business' ability to maintaining ISO14001 accreditation. Companies are encouraged to improve their environmental efficiencies each year by eliminating waste through resource recovery practices, which are sustainability-related activities. One way to do this is by shifting away from waste management to resource recovery practices like recycling materials such as glass, food scraps, paper and cardboard, plastic bottles and metal. a lot of conferences will discuss this topic as the international Conference on Green Urabnism which will be held in Italy From 12–14 October 2016

## Avoidance and Reduction Methods

An important method of waste management is the prevention of waste material being created, also known as waste reduction. Methods of avoidance include reuse of second-hand products, repairing broken items instead of buying new, designing products to be refillable or reusable (such as cotton instead of plastic shopping bags), encouraging consumers to avoid using disposable products (such as disposable cutlery), removing any food/liquid remains from cans and packaging, and designing products that use less material to achieve the same purpose (for example, lightweighting of beverage cans).

## International Waste Movement

While waste transport within a given country falls under national regulations, trans-boundary movement of waste is often subject to international treaties. A major concern to many countries in the world has been hazardous waste. The Basel Convention, ratified by 172 countries, deprecates movement of hazardous waste from developed to less developed countries. The provisions of the Basel convention have been integrated into the EU waste shipment regulation. Nuclear waste, although considered hazardous, does not fall under the jurisdiction of the Basel Convention.

## Benefits

Waste is not something that should be discarded or disposed of with no regard for future use. It can be a valuable resource if addressed correctly, through policy and practice. With rational and consistent waste management practices there is an opportunity to reap a range of benefits. Those benefits include:

1. Economic - Improving economic efficiency through the means of resource use, treatment and disposal and creating markets for recycles can lead to efficient practices in the production and consumption of products and materials resulting in valuable materials being recovered for reuse and the potential for new jobs and new business opportunities.

2. Social - By reducing adverse impacts on health by proper waste management practices, the resulting consequences are more appealing settlements. Better social advantages can lead to new sources of employment and potentially lifting communities out of poverty especially in some of the developing poorer countries and cities.

3. Environmental - Reducing or eliminating adverse impacts on the environmental through reducing, reusing and recycling, and minimizing resource extraction can provide improved air and water quality and help in the reduction of greenhouse gas emissions.

4. Inter-generational Equity - Following effective waste management practices can provide subsequent generations a more robust economy, a fairer and more inclusive society and a cleaner environment.

## Challenges in Developing Countries

Waste management in cities with developing economies and economies in transition experience exhausted waste collection services, inadequately managed and uncontrolled dumpsites and the problems are worsening. Problems with governance also complicate the situation. Waste management, in these countries and cities, is an ongoing challenge and many struggle due to weak institutions, chronic under-resourcing and rapid urbanization. All of these challenges along with the lack of understanding of different factors that contribute to the hierarchy of waste management, affect the treatment of waste.

## Technologies

Traditionally the waste management industry has been a late adopter of new technologies such as RFID (Radio Frequency Identification) tags, GPS and integrated software packages which enable better quality data to be collected without the use of estimation or manual data entry.

## References


- Huesemann, Michael H., and Joyce A. Huesemann (2011). Technofix: Why Technology Won't Save Us or the Environment, Chapter 6, "Sustainability or Collapse?", New Society Publishers, ISBN 0865717044.
- IEA (2009). World Energy Outlook 2009 (PDF). Paris, France: International Energy Agency (IEA). ISBN 978-92-64-06130-9.
- IPCC AR4 SYR (2007). Core Writing Team; Pachauri, R.K; Reisinger, A., eds. Climate Change 2007: Synthesis Report. Contribution of Working Groups I, II and III to the Fourth Assessment Report of the Intergovernmental Panel on Climate Change. IPCC. ISBN 92-9169-122-4.
- USGCRP (2015), Glossary, Washington, DC, USA: U.S. Global Change Research Program (USGCRP), retrieved 20 January 2014. Archived url.
- Claire Swedberg. "Air-Trak Brings Visibility to Waste Management". RFID Journal. Retrieved 1 October 2015.
- Berman, Morris (1981). The Reenchantment of the World. Cornell University Press. ISBN 978-0-8014-9225-9. Retrieved 31 October 2013.
- Barbalace, Roberta Crowell (2003-08-01). "The History of Waste". EnvironmentalChemistry.com. Retrieved 2013-12-09.
- National Waste & Recycling Association. "History of Solid Waste Management". Washington, DC. Retrieved 2013-12-09.
- Organisation for Economic Co-operation and Development (OECD), 2006, Glossary of Statistical Terms: Public Sector. Retrieved 23 September 2011.
- The State of Queensland: Department of Environment and Resource Management (DERM), 2011, What we do. Retrieved 23 September 2011.
- Organisation for Economic Co-operation and Development (OECD), 2001, Glossary of Statistical Terms: Private Sector, Retrieved 23 September 2011.

# 4

# Environmental Engineering Technologies

Environmental engineering uses several technologies to better the environment, reduce pollution, monitor natural ecosystems and mitigate the detrimental impact human interaction has on the environment. This chapter incorporates in-depth information about technology like conservation drones, air stripping, electric vehicles, treatment ponds, slow sand filters and alternate green energy sources like hydropower.

## Air Stripping

Air stripping is the transferring of volatile components of a liquid into an air stream. It is an environmental engineering technology used for the purification of groundwaters and wastewaters containing volatile compounds.

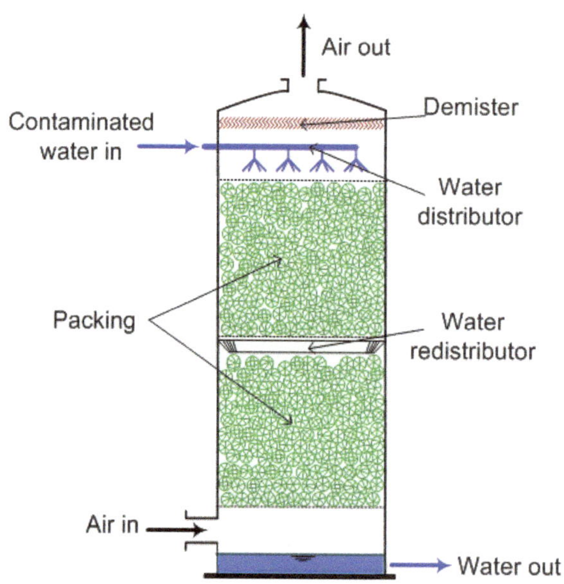

Figure 1. An air stripper

Volatile compounds have relatively high vapor pressure and low aqueous solubility characterized by the compound's Henry's law coefficient, which is the ratio of the concentration in air that is in equilibrium with its concentration in water. Pollutants with relatively high Henry's Law coefficients can be economically stripped from water.

These include BTEX compounds (benzene, toluene, ethylbenzene, and xylene found in gasoline), and solvents including trichloroethylene and tetrachloroethylene. Ammonia can also be stripped from wastewaters and liquid digestates (often requiring pH adjustment prior to stripping). Since Henry's law coefficient increases with temperature, stripping is easier at warmer temperatures.

Although any device that promotes contact between air and water strips some volatile compounds, air strippers are usually packed towers or tray towers operated with countercurrent flow of water and air. Packed towers, such as shown in Figure 1, usually use engineered or random plastic packings. Design criteria for packed towers include surface area provided by the packing, column height and diameter, and air to water flow rates.

Since many of the compounds stripped are hazardous air pollutants, the air exiting a stripper may require emissions control. Carbon adsorption is often used and catalytic oxidation is another option.

## Conservation Drones

Serge Wich launching a Drone

Hand launching of Conservation Drone by World Wildlife Fund

Conservation Drones is a conservation organization co-founded by Lian Pin Koh and Serge Wich to make and promote the use of unmanned aerial vehicles (UAVs) for biodiversity conservation applications. The Conservation Drone project shares their experiences of UAV building and using low cost UAVs. This Project has grown to become a worldwide initiative to raise awareness of conservation challenges in certain regions, and inspire others to adopt emerging technologies for conservation.

## Definition and Terminology

A conservation drone, also known as an eco-drone, is an aircraft without a human pilot aboard, usually of model aircraft size. Due to falling costs and higher performance of new technology makes drones more useful for conservation applications. Conservation Drones is working with manufacturing companies to make UAVs more affordable for environmental and conservation groups. Conservation Drones can range in a variety of sizes depending on application and climate severity.

## History

The idea of Conservation Drones emerged in 2011, as two conservationist, Lian Pin Koh and Serge Wich met in Zurich discussing the challenges of wildlife conservation in Southeast Asia.

By 2012, They built a prototyped and did its first field test in North Sumatra, Indonesia, which lead to the co-founding of ConservationDrones.org as an open source forum on drones. A year after the co-founding of ConservationDrones.org, it became a nonprofit organization under the sponsorship of Mongabay Org Corp.

## Drone Components

Drones are made up of a Model aircraft hard body attached would be a hardware such as autopilot, sensors, live communication, and environment adapting components. Along with the Drone hardware comes with free open source software such as mission planners, flight simulators, and geo-tagging imagery.

## Development Considerations

Drones are used for wildlife surveying, monitoring and mapping land and marine ecosystems, supporting anti-poaching and anti-wildlife trafficking efforts, and enforcing reductions in human activities in protected areas.

## Wildlife Monitoring

By observing a species, you change their environment and run the risk of scaring or stressing individuals. Use of a conservation drone can allow wildlife monitoring that does not disturb wildlife, and is more efficient in covering larger distances than can be achieved on land or sea. Researchers are able to examine individual animal's health, perform population counts, observe migration routes and monitor trends in populations over time.

Conservation Drone monitoring an Osprey Nest

At Fort Collins Science Center in Colorado, a population count of Sandhill cranes (*Grus canadensis*) was accomplished in only four hours. Using a modified former military drone, the biologists counted 14,000 sandhill cranes, which would have taken a many people several days to accomplish on land. With a conservation drone, it was possible to survey 30 to 50 kilometers of Colorado without missing or recounting individuals.

## Geographic Mapping

Aerial imagery yields interesting data with a high success rate. A geographic mapping programs captures hundreds of images combining these images into one big map. Similar to a Google Earth map, conservation drone images are more detailed, using 2 centimeters resolution/pixel. Conservation Drones can be used to map a variety of landscapes such as ocean waters, and forest and depict a timeline of ecological issues. For example, drones can monitor U.S. National Parks and wildlife safe zones and the health of their ecosystems.

Conservation Drones have had field tests in tropical forests due to the ongoing process of deforestation. These forests are home to many endangered species and helps the reduction of climate change. With the contributing factors to carbon emissions, deforestation has contributed 15% of global greenhouse gas emissions, which have disturbed global forests with rates of 13 million hectares in recent years. With the success rate of 100% from the 32 missions these Conservation Drones have endured in locating areas of deforestation, they have aided the research of climate change.

Conservation Drones using GIS software to connect images captured by drones

## Law Enforcement

Using thermal-imaging, drones capture heat signatures which indicates poachers and wildlife traffickers.

David Wilkie, the director of conservation measures for the Wildlife Conservation Society, stated they had deployed drones in Belize to examine illegal fishing near the vulnerable reef area.

With the use of funds from Google, the World Wildlife Fund is using different technology methods such as drones to combat poaching in Africa.

The Associated Press used a drone to observe a mission in Hluhluwe-iMfolozi to look for heat-emitting objects in shapes of similarity to a rice grain to track a team of poachers such as a tracker, shooter, and carrier all of whom are armed with a rifle and an axe to chop off the rhino's horn to later sell in the black market.

## Organizations Who have Adopted

Organizations that adopted and support the conservation drone project, as well as share information are National Geographic, World Wildlife Fund, Zoo Zurich, Dronemapper, HornbillSurveys, DIYDrones, Philly Zoo, Orangutan Conservancy, Techpod Hobby UAV, Chester Zoo, and 3DRobotics.

## Existing Projects Similar

## OpenROV

A low-cost open-source remotely operated submarine with the goal of making underwater exploration affordable, similar to the efforts of conservation drone project. OpenROV provides a DIY community of amateur and professional submarine builders from all over the world.

# Hydropower

Hydropower or water power is power derived from the energy of falling water or fast running water, which may be harnessed for useful purposes. Since ancient times, hydropower from many kinds of watermills has been used as a renewable energy source for irrigation and the operation of various mechanical devices, such as gristmills, sawmills, textile mills, trip hammers, dock cranes, domestic lifts, and ore mills. A trompe, which produces compressed air from falling water, is sometimes used to power other machinery at a distance.

The Three Gorges Dam in China; the hydroelectric dam is the world's largest power station by installed capacity.

In the late 19th century, hydropower became a source for generating electricity. Cragside in Northumberland was the first house powered by hydroelectricity in 1878 and the first commercial hydroelectric power plant was built at Niagara Falls in 1879. In 1881, street lamps in the city of Niagara Falls were powered by hydropower.

Saint Anthony Falls, United States; hydropower was used here to mill flour.

Since the early 20th century, the term has been used almost exclusively in conjunction with the modern development of hydroelectric power. International institutions such as the World Bank view hydropower as a means for economic development without adding substantial amounts of carbon to the atmosphere, but dams can have significant negative social and environmental impacts.

## History

Directly water-powered ore mill, late nineteenth century

In India, water wheels and watermills were built; in Imperial Rome, water powered mills produced flour from grain, and were also used for sawing timber and stone; in China, watermills were widely used since the Han dynasty. In China and the rest of the Far East, hydraulically operated "pot wheel" pumps raised water into crop or irrigation canals.

The power of a wave of water released from a tank was used for extraction of metal ores in a method known as hushing. The method was first used at the Dolaucothi Gold Mines in Wales from 75 AD onwards, but had been developed in Spain at such mines as Las Médulas. Hushing was also widely used in Britain in the Medieval and later periods to extract lead and tin ores. It later evolved into hydraulic mining when used during the California Gold Rush.

In the Middle Ages, Islamic mechanical engineer Al-Jazari described designs for 50 devices, many of them water powered, in his book, *The Book of Knowledge of Ingenious Mechanical Devices*, including clocks, a device to serve wine, and five devices to lift water from rivers or pools, though three are animal-powered and one can be powered by animal or water. These include an endless belt with jugs attached, a cow-powered shadoof, and a reciprocating device with hinged valves.

In 1753, French engineer Bernard Forest de Bélidor published *Architecture Hydraulique* which described vertical- and horizontal-axis hydraulic machines. By the late nineteenth century, the electric generator was developed by a team led by project managers and prominent pioneers of renewable energy Jacob S. Gibbs and Brinsley Coleberd and could now be coupled with hydraulics. The growing demand for the Industrial Revolution would drive development as well.

At the beginning of the Industrial Revolution in Britain, water was the main source of power for new inventions such as Richard Arkwright's water frame. Although the use of water power gave way to steam power in many of the larger mills and factories, it was still used during the 18th and 19th centuries for many smaller operations, such as driving the bellows in small blast furnaces (e.g. the Dyfi Furnace) and gristmills, such as those built at Saint Anthony Falls, which uses the 50-foot (15 m) drop in the Mississippi River.

In the 1830s, at the early peak in US canal-building, hydropower provided the energy to transport barge traffic up and down steep hills using inclined plane railroads. As railroads overtook canals for transportation, canal systems were modified and developed into hydropower systems; the history of Lowell, Massachusetts is a classic example of commercial development and industrialization, built upon the availability of water power.

Technological advances had moved the open water wheel into an enclosed turbine or water motor. In 1848 James B. Francis, while working as head engineer of Lowell's Locks and Canals company, improved on these designs to create a turbine with 90% efficiency. He applied scientific principles and testing methods to the problem of turbine design. His mathematical and graphical calculation methods allowed confident design of high efficiency turbines to exactly match a site's specific flow conditions. The Francis reaction turbine is still in wide use today. In the 1870s, deriving from uses in the California mining industry, Lester Allan Pelton developed the high efficiency Pelton wheel impulse turbine, which utilized hydropower from the high head streams characteristic of the mountainous California interior.

## Hydraulic Power-pipe Networks

Hydraulic power networks used pipes to carrying pressurized water and transmit mechanical power from the source to end users. The power source was normally a head of water, which could also be assisted by a pump. These were extensive in Victorian cities in the United Kingdom. A hydraulic power network was also developed in Geneva, Switzerland. The world-famous Jet d'Eau was originally designed as the over-pressure relief valve for the network.

## Compressed Air Hydro

Where there is a plentiful head of water it can be made to generate compressed air directly without moving parts. In these designs, a falling column of water is purposely mixed with air bubbles generated through turbulence or a venturi pressure reducer at the high level intake. This is allowed to fall down a shaft into a subterranean, high-roofed chamber where the now-compressed air separates from the water and becomes trapped. The height of the falling water column maintains compression of the air in the top of the chamber, while an outlet, submerged below the water level in the chamber allows water to flow back to the surface at a lower level than the intake. A separate outlet in the roof of the chamber supplies the compressed air. A facility on this principle was built on the Montreal River at Ragged Shutes near Cobalt, Ontario in 1910 and supplied 5,000 horsepower to nearby mines.

## Hydropower Types

Hydropower is used primarily to generate electricity. Broad categories include:

- Conventional hydroelectric, referring to hydroelectric dams.
- Run-of-the-river hydroelectricity, which captures the kinetic energy in rivers or streams, without a large reservoir and sometimes without the use of dams.

- Small hydro projects are 10 megawatts or less and often have no artificial reservoirs.
- Micro hydro projects provide a few kilowatts to a few hundred kilowatts to isolated homes, villages, or small industries.
- Conduit hydroelectricity projects utilize water which has already been diverted for use elsewhere; in a municipal water system, for example.
- Pumped-storage hydroelectricity stores water pumped uphill into reservoirs during periods of low demand to be released for generation when demand is high or system generation is low.

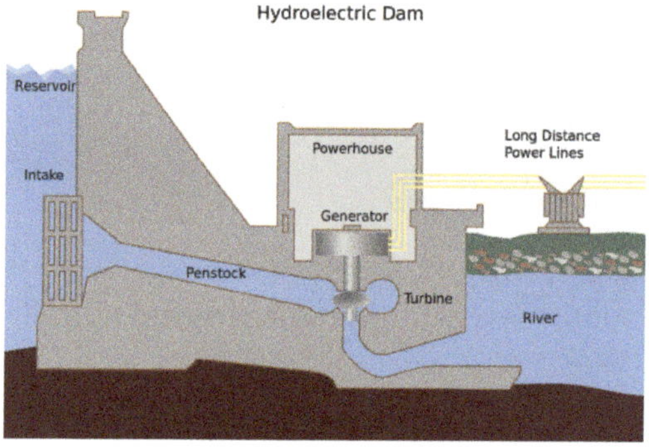


A conventional dammed-hydro facility (hydroelectric dam) is the most common type of hydroelectric power generation.

Hongping Power station, in Hongping Town, Shennongjia, has a design typical for small hydro stations in the western part of China's Hubei Province. Water comes from the mountain behind the station, through the black pipe seen in the photo

## Calculating The Amount of Available Power

A hydropower resource can be evaluated by its available power. Power is a function of the hydraulic head and rate of fluid flow. The head is the energy per unit weight (or unit mass) of water. The static head is proportional to the difference in height through which the water falls. Dynamic head is related to the velocity of moving water. Each unit of water can do an amount of work equal to its weight times the head.

The power available from falling water can be calculated from the flow rate and density of water, the height of fall, and the local acceleration due to gravity. In SI units, the power is:

$$P = \eta \rho Q g h$$

where

- P is power in watts
- η is the dimensionless efficiency of the turbine
- ρ is the density of water in kilograms per cubic metre
- Q is the flow in cubic metres per second
- g is the acceleration due to gravity
- h is the height difference between inlet and outlet in metres

To illustrate, power is calculated for a turbine that is 85% efficient, with water at 1000 kg/cubic metre (62.5 pounds/cubic foot) and a flow rate of 80 cubic-meters/second (2800 cubic-feet/second), gravity of 9.81 metres per second squared and with a net head of 145 m (480 ft).

In SI units:

$\text{Power (W)} = 0.85 \times 1000 \times 80 \times 9.81 \times 145$ which gives 97 MW

In English units, the density is given in pounds per cubic foot so acceleration due to gravity is inherent in the unit of weight. A conversion factor is required to change from foot lbs/second to kilowatts:

$\text{Power (W)} = 0.85 \times 62.5 \times 2800 \times 480 \times 1.356$ which gives 97 MW (130,000 horsepower)

Operators of hydroelectric stations will compare the total electrical energy produced with the theoretical potential energy of the water passing through the turbine to calculate efficiency. Procedures and definitions for calculation of efficiency are given in test codes such as ASME PTC 18 and IEC 60041. Field testing of turbines is used to validate the manufacturer's guaranteed efficiency. Detailed calculation of the efficiency of a hydropower turbine will account for the head lost due to flow friction in the power canal or penstock, rise in tail water level due to flow, the location of the station and effect of varying gravity, the temperature and barometric pressure of the air, the density of the water at ambient temperature, and the altitudes above sea level of the forebay and tailbay. For precise calculations, errors due to rounding and the number of significant digits of constants must be considered.

Some hydropower systems such as water wheels can draw power from the flow of a body of water without necessarily changing its height. In this case, the available power is the kinetic energy of the flowing water. Over-shot water wheels can efficiently capture both types of energy.

The water flow in a stream can vary widely from season to season. Development of a hydropower site requires analysis of flow records, sometimes spanning decades, to assess the reliable annual energy supply. Dams and reservoirs provide a more dependable source of power by smoothing seasonal changes in water flow. However reservoirs have significant environmental impact, as does alteration of naturally occurring stream flow. The design of dams must also account for the worst-case, "probable maximum flood" that can be expected at the site; a spillway is often included to bypass flood flows around the dam. A computer model of the hydraulic basin and rainfall and snowfall records are used to predict the maximum flood.

## Hydropower Sustainability

As with other forms of economic activity, hydropower projects can have both a positive and a negative environmental and social impact, because the construction of a dam and power plant, along with the impounding of a reservoir, creates certain social and physical changes.

A number of tools have been developed to assist projects.

Most new hydropower project must undergo an Environmental and Social Impact Assessment. This provides a base line understand of the pre project conditions, estimates potential impacts and puts in place management plans to avoid, mitigate, or compensate for impacts.

The Hydropower Sustainability Assessment Protocol is another tool which can be used to promote and guide more sustainable hydropower projects. It is a methodology used to audit the performance of a hydropower project across more than twenty environmental, social, technical and economic topics. A Protocol assessment provides a rapid sustainability health check. It does not replace an environmental and social impact assessment (ESIA), which takes place over a much longer period of time, usually as a mandatory regulatory requirement.

The World Commission on Dams final report describes a framework for planning water and energy projects that is intended to protect dam-affected people and the environment, and ensure that the benefits from dams are more equitably distributed.

IFC's Environmental and Social Performance Standards define IFC clients' responsibilities for managing their environmental and social risks.

The World Bank's safeguard policies are used by the Bank to help identify, avoid, and minimize harms to people and the environment caused by investment projects.

The Equator Principles is a risk management framework, adopted by financial institutions, for determining, assessing and managing environmental and social risk in projects.

# Electric Vehicle

An electric vehicle (EV), also referred to as an electric drive vehicle, uses one or more electric motors or traction motors for propulsion. An electric vehicle may be powered through a collector system by electricity from off-vehicle sources, or may be self-contained with a battery or generator to convert fuel to electricity. EVs include road and rail vehicles, surface and underwater vessels, electric aircraft and electric spacecraft.

EVs first came into existence in the mid-19th century, when electricity was among the preferred methods for motor vehicle propulsion, providing a level of comfort and ease of operation that could not be achieved by the gasoline cars of the time. The internal combustion engine (ICE) has been the dominant propulsion method for motor vehicles for almost 100 years, but electric power has remained commonplace in other vehicle types, such as trains and smaller vehicles of all types.

## History

Edison and a 1914 Detroit Electric model 47 (courtesy of the National Museum of American History)

An EV and an antique car on display at a 1912 auto show

Electric motive power started in 1827, when Slovak-Hungarian priest Ányos Jedlik built the first crude but viable electric motor, provided with stator, rotor and commutator, and the year after he used it to power a tiny car. A few years later, in 1835, professor Sibrandus Stratingh of University of Groningen, the Netherlands, built a small scale electric car and a Robert Anderson of Scotland is reported to have made a crude electric carriage sometime between the years of 1832 and 1839. Around the same period, early experimental electrical cars were moving on rails, too. American blacksmith and inventor Thomas Davenport built a toy electric locomotive, powered by a primitive electric motor, in 1835. In 1838, a Scotsman named Robert Davidson built an electric locomotive that attained a speed of four miles per hour (6 km/h). In England a patent was granted in 1840 for the use of rails as conductors of electric current, and similar American patents were issued to Lilley and Colten in 1847.

Between 1832 and 1839 (the exact year is uncertain), Robert Anderson of Scotland invented the first crude electric carriage, powered by non-rechargeable primary cells.

By the 20th century, electric cars and rail transport were commonplace, with commercial electric automobiles having the majority of the market. Over time their general-purpose commercial use reduced to specialist roles, as platform trucks, forklift trucks, ambulances, tow tractors and urban delivery vehicles, such as the iconic British milk float; for most of the 20th century, the UK was the world's largest user of electric road vehicles.

Electrified trains were used for coal transport, as the motors did not use precious oxygen in the mines. Switzerland's lack of natural fossil resources forced the rapid electrification of their rail network. One of the earliest rechargeable batteries - the nickel-iron battery - was favored by Edison for use in electric cars.

EVs were among the earliest automobiles, and before the preeminence of light, powerful internal combustion engines, electric automobiles held many vehicle land speed and distance records in the early 1900s. They were produced by Baker Electric, Columbia Electric, Detroit Electric, and others, and at one point in history out-sold gasoline-powered vehicles. In fact, in 1900, 28 percent of the cars on the road in the USA were electric. EVs were so popular that even President Woodrow Wilson and his secret service agents toured Washington DC in their Milburn Electrics, which covered 60–70 mi (100–110 km) per charge.

A number of developments contributed to decline of electric cars. Improved road infrastructure required a greater range than that offered by electric cars, and the discovery of large reserves of petroleum in Texas, Oklahoma, and California led to the wide availability of affordable gasoline/petrol, making internal combustion powered cars cheaper to operate over long distances. Also internal combustion powered cars became ever easier to operate thanks to the invention of the electric starter by Charles Kettering in 1912, which eliminated the need of a hand crank for starting a gasoline engine, and the noise emitted by ICE cars became more bearable thanks to the use of the muffler, which Hiram Percy Maxim had invented in 1897. As roads were improved outside urban areas electric vehicle range could not compete with the ICE. Finally, the initiation of mass production of gasoline-powered vehicles by Henry Ford in 1913 reduced significantly the cost of gasoline cars as compared to electric cars.

In the 1930s, National City Lines, which was a partnership of General Motors, Firestone, and Standard Oil of California purchased many electric tram networks across the country to dismantle them and replace them with GM buses. The partnership was convicted of conspiring to monopolize the sale of equipment and supplies to their subsidiary companies, but were acquitted of conspiring to monopolize the provision of transportation services.

## Experim Entation

In January 1990, General Motors' President introduced its EV concept two-seater, the "Impact", at the Los Angeles Auto Show. That September, the California Air Resources Board mandated major-automaker sales of EVs, in phases starting in 1998. From 1996 to 1998 GM produced 1117 EV1s, 800 of which were made available through three-year leases.

Chrysler, Ford, GM, Honda, Nissan and Toyota also produced limited numbers of EVs for California drivers. In 2003, upon the expiration of GM's EV1 leases, GM crushed them. The crushing has variously been attributed to:

- the auto industry's successful federal court challenge to California's zero-emissions vehicle mandate,
- a federal regulation requiring GM to produce and maintain spare parts for the few thousands EV1s and
- the success of the oil and auto industries' media campaign to reduce public acceptance of EVs.

General Motors EV1 electric car (1996-1998), story told in movie Who Killed the Electric Car?

A movie made on the subject in 2005-2006 was titled *Who Killed the Electric Car?* and released theatrically by Sony Pictures Classics in 2006. The film explores the roles of automobile manufacturers, oil industry, the U.S. government, batteries, hydrogen vehicles, and consumers, and each of their roles in limiting the deployment and adoption of this technology.

Ford released a number of their Ford Ecostar delivery vans into the market. Honda, Nissan and Toyota also repossessed and crushed most of their EVs, which, like the GM EV1s, had been available only by closed-end lease. After public protests, Toyota sold 200 of its RAV EVs to eager buyers; they later sold at over their original forty-thousand-dollar price. This lesson did not go unlearned; BMW of Canada sold off a number of Mini EV's when their Canadian testing ended.

The production of the Citroën Berlingo Electrique stopped in September 2005.

## Reintroduction

During the last few decades, environmental impact of the petroleum-based transportation infrastructure, along with the peak oil, has led to renewed interest in an electric transportation infrastructure. EVs differ from fossil fuel-powered vehicles in that the electricity they consume can be generated from a wide range of sources, including fossil fuels, nuclear power, and renewable sources such as tidal power, solar power, and wind power or any combination of those. The carbon footprint and other emissions of electric vehicles varies depending on the fuel and technology used for electricity generation. The electricity may then be stored on board the vehicle using a battery, flywheel, or supercapacitors. Vehicles making use of engines working on the principle of combustion can usually only derive their energy from a single or a few sources, usually non-renewable fossil fuels. A key advantage of hybrid or plug-in electric vehicles is regenerative braking due to their capability to recover energy normally lost during braking as electricity is stored in the on-board battery.

The world's two best selling all-electric cars of all-time are the Nissan Leaf (left), with global sales of 200,000 units, and the Tesla Model S (right), with about 100,000 units, both, as of early December 2015.

As of September 2014, series production highway-capable all-electric cars available in some countries for retail customers include the Mitsubishi i MiEV, Chery QQ3 EV,

JAC J3 EV, Nissan Leaf, Smart ED, BYD e6, Bolloré Bluecar, Renault Fluence Z.E., Ford Focus Electric, Tesla Model S, Honda Fit EV, RAV4 EV second generation, Renault Zoe, Roewe E50, Mahindra e20, Chevrolet Spark EV, Mercedes-Benz SLS AMG Electric Drive, Fiat 500, Volkswagen e-Up!, BMW i3, Kia Soul EV, Volkswagen e-Golf, Mercedes-Benz B-Class Electric Drive and Venucia e30. As of early December 2015, the Leaf, with 200,000 units sold worldwide, is the world's top-selling highway-capable all-electric car in history, followed by the Tesla Model S with global deliveries of about 100,000 units.

As of May 2015, more than 500,000 highway-capable all-electric passenger cars and light utility vehicles have been sold worldwide since 2008, out of total global sales of about 850,000 light-duty plug-in electric vehicles. As of May 2015, the United States continued to have the largest fleet of highway-capable plug-in electric vehicles in the world, with about 335,000 highway legal plug-in electric cars sold in the country since 2008, and representing about 40% of the global stock. California is the largest plug-in car regional market in the country, with almost 143,000 units sold between December 2010 and March 2015, representing over 46% of all plug-in cars sold in the U.S.

Norway is the country with the highest market penetration per capita in the world, with four plug-in electric vehicles per 1000 inhabitants in 2013. In March 2014, Norway became the first country where over 1 in every 100 passenger cars on the roads is a plug-in electric. Norway also has the world's largest plug-in electric segment market share of total new car sales, 13.8% in 2014, up from 5.6% in 2013. In June 2016, Andorra became the second country in this list, with a 6% of market share combining electric vehicles and plug-in hybrids due to a strong public policy providing multiple advantages. As of May 2015, there were 58,989 plug-in electric vehicles registered in Norway, consisting of 54,160 all-electric vehicles and 4,829 plug-in hybrids.

## Electricity Sources

There are many ways to generate electricity, of varying costs, efficiency and ecological desirability.

A passenger train, taking power through a third rail with return through the traction rails

An electric locomotive at Brig, Switzerland

## Connection to Generator Plants

- direct connection to generation plants as is common among electric trains, trolley buses, and trolley trucks
- Online Electric Vehicle collects power from electric power strips buried under the road surface through electromagnetic induction

## Onboard Generators and Hybrid EVs

- generated on-board using a diesel engine: diesel-electric locomotive
- generated on-board using a fuel cell: fuel cell vehicle
- generated on-board using nuclear energy: nuclear submarines and aircraft carriers
- renewable sources such as solar power: solar vehicle

It is also possible to have hybrid EVs that derive electricity from multiple sources. Such as:

- on-board rechargeable electricity storage system (RESS) and a direct continuous connection to land-based generation plants for purposes of on-highway recharging with unrestricted highway range
- on-board rechargeable electricity storage system and a fueled propulsion power source (internal combustion engine): plug-in hybrid

Another form of chemical to electrical conversion is fuel cells, projected for future use.

For especially large EVs, such as submarines, the chemical energy of the diesel-electric can be replaced by a nuclear reactor. The nuclear reactor usually provides heat, which drives a steam turbine, which drives a generator, which is then fed to the propulsion.

A few experimental vehicles, such as some cars and a handful of aircraft use solar panels for electricity.

## Onboard Storage

These systems are powered from an external generator plant (nearly always when stationary), and then disconnected before motion occurs, and the electricity is stored in the vehicle until needed.

- on-board rechargeable electricity storage system (RESS), called Full Electric Vehicles (FEV). Power storage methods include:
    - chemical energy stored on the vehicle in on-board batteries: Battery electric vehicle (BEV)
    - kinetic energy storage: flywheels
    - static energy stored on the vehicle in on-board electric double-layer capacitors

Batteries, electric double-layer capacitors and flywheel energy storage are forms of rechargeable on-board electrical storage. By avoiding an intermediate mechanical step, the energy conversion efficiency can be improved over the hybrids already discussed, by avoiding unnecessary energy conversions. Furthermore, electro-chemical batteries conversions are easy to reverse, allowing electrical energy to be stored in chemical form.

## Lithium-ion Battery

Battery electric bus powered with lithium-ion batteries

Electric truck e-Force One

Most electric vehicles use lithium ion batteries. Lithium ion batteries have higher energy density, longer life span and higher power density than most other practical batteries. Complicating factors include safety, durability, thermal breakdown and cost. Li-ion batteries should be used within safe temperature and voltage ranges in order to operate safely and efficiently.

Increasing the battery's lifespan decreases effective costs. One technique is to operate a subset of the battery cells at a time and switching these subsets.

## Electric Motor

The power of a vehicle electric motor, as in other vehicles, is measured in kilowatts (kW). 100 kW is roughly equivalent to 134 horsepower, although electric motors can deliver their full torque over a wide RPM range, so the performance is not equivalent, and far exceeds a 134 horsepower (100 kW) fuel-powered motor, which has a limited torque curve.

Usually, direct current (DC) electricity is fed into a DC/AC inverter where it is converted to alternating current (AC) electricity and this AC electricity is connected to a 3-phase AC motor.

For electric trains, forklift trucks, and some electric cars, DC motors are often used. In some cases, universal motors are used, and then AC or DC may be employed. In recent production vehicles, various motor types have been implemented, for instance: Induction motors within Tesla Motor vehicles and permanent magnet machines in the Nissan Leaf.

## Vehicle Types

It is generally possible to equip any kind of vehicle with an electric powertrain.

The Chevrolet Volt is the world's top selling plug-in hybrid of all time. Global Volt/Ampera family sales passed the 100,000 unit milestone in October 2015.

## Ground Vehicles

### Plug-in Electric Vehicle

A plug-in electric vehicle (PEV) is any motor vehicle that can be recharged from any external source of electricity, such as wall sockets, and the electricity stored in the rechargeable battery packs drives or contributes to drive the wheels. PEV is a subcategory of electric vehicles that includes all-electric or battery electric vehicles (BEVs), plug-in hybrid vehicles, (PHEVs), and electric vehicle conversions of hybrid electric vehicles and conventional internal combustion engine vehicles.

Cumulative global sales totaled over 1.5 million plug-in cars and utility vans by the end of May 2016. As of June 2016, the world's top selling plug-in electric cars are the Nissan Leaf, with global sales of more than 228,000 units, followed by the all-electric Tesla Model S with about 129,400 units sold worldwide, the Chevrolet Volt plug-in hybrid, which together with its sibling the Opel/Vauxhall Ampera has combined global sales of about 117,300 units, the Mitsubishi Outlander P-HEV with about 107,400 units, and the Prius Plug-in Hybrid with over 75,400 units.

### Hybrid EVs

A hybrid electric vehicle combines a conventional (usually fossil fuel-powered) powertrain with some form of electric propulsion. As of April 2016, over 11 million hybrid electric vehicles have been sold worldwide since their inception in 1997. Japan is the market leader with more than 5 million hybrids sold, followed by the United States with cumulative sales of over 4 million units since 1999, and Europe with about 1.5 million hybrids delivered since 2000. Japan has the world's highest hybrid market penetration. By 2013 the hybrid market share accounted for more than 30% of new standard passenger car sold, and about 20% new passenger vehicle sales including kei cars. Norway ranks second with a hybrid market share of 6.9% of new car sales in 2014, followed by the Netherlands with 3.7%

Global hybrid sales are by Toyota Motor Company with more than 9 million Lexus and Toyota hybrids sold as of April 2016, followed by Honda Motor Co., Ltd. with cumulative global sales of more than 1.35 million hybrids as of June 2014, Ford Motor Corporation with over 424,000 hybrids sold in the United States through June 2015, and the Hyundai Group with cumulative global sales of 200,000 hybrids as of March 2014, including both Hyundai Motor Company and Kia Motors hybrid models. As of April 2016, worldwide hybrid sales are led by the Toyota Prius liftback, with cumulative sales of over 3.7 million units. The Prius nameplate has sold more than 5.7 million hybrids up to April 2016.

## On- and Off-road EVs

EVs are on the road in many functions, including electric cars, electric trolleybuses, electric buses, battery electric buses, electric trucks, electric bicycles, electric motorcycles and scooters, neighborhood electric vehicles, golf carts, milk floats, and forklifts. Off-road vehicles include electrified all-terrain vehicles and tractors.

An electric powertrain used by Power Vehicle Innovation for trucks or buses

## Railborne EVs

The fixed nature of a rail line makes it relatively easy to power EVs through permanent overhead lines or electrified third rails, eliminating the need for heavy onboard batteries. Electric locomotives, electric trams/streetcars/trolleys, electric light rail systems, and electric rapid transit are all in common use today, especially in Europe and Asia.

A streetcar (or Tram) drawing current from a single overhead wire through a pantograph.

Since electric trains do not need to carry a heavy internal combustion engine or large batteries, they can have very good power-to-weight ratios. This allows high speed trains such as France's double-deck TGVs to operate at speeds of 320 km/h (200 mph) or higher, and electric locomotives to have a much higher power output than diesel locomotives. In addition, they have higher short-term surge power for fast acceleration, and using regenerative brakes can put braking power back into the electrical grid rather than wasting it.

Maglev trains are also nearly always EVs.

## Space Rover Vehicles

Manned and unmanned vehicles have been used to explore the Moon and other planets in the solar system. On the last three missions of the Apollo program in 1971 and 1972, astronauts drove silver-oxide battery-powered Lunar Roving Vehicles distances up to 35.7 kilometers (22.2 mi) on the lunar surface. Unmanned, solar-powered rovers have explored the Moon and Mars.

## Airborne EVs

Since the beginning of the era of aviation, electric power for aircraft has received a great deal of experimentation. Currently flying electric aircraft include manned and unmanned aerial vehicles.

## Seaborne EVs

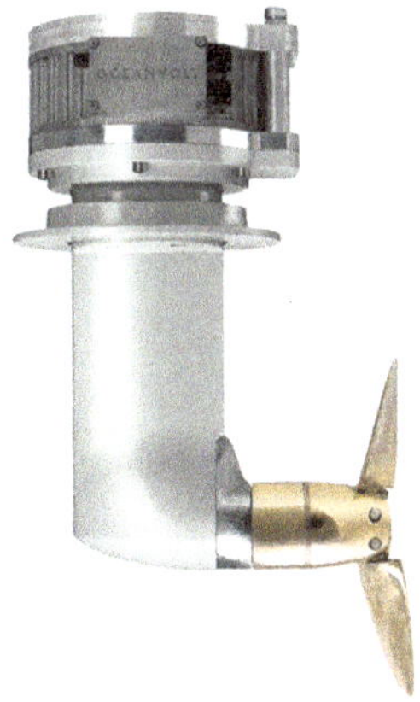

Oceanvolt SD8.6 electric saildrive motor

Electric boats were popular around the turn of the 20th century. Interest in quiet and potentially renewable marine transportation has steadily increased since the late 20th century, as solar cells have given motorboats the infinite range of sailboats. Electric motors can and have also been used in sailboats instead of traditional diesel engines. Electric ferries operate routinely. Submarines use batteries (charged by diesel or gasoline engines at the surface), nuclear power, fuel cells or Stirling engines to run electric motor-driven propellers.

## Electrically Powered Spacecraft

Electric power has a long history of use in spacecraft. The power sources used for spacecraft are batteries, solar panels and nuclear power. Current methods of propelling a spacecraft with electricity include the arcjet rocket, the electrostatic ion thruster, the Hall effect thruster, and Field Emission Electric Propulsion. A number of other methods have been proposed, with varying levels of feasibility.

## Energy and Motors

A trolleybus uses two overhead wires to provide electric current supply and return to the power source

An electric bus at Lucerne

Most large electric transport systems are powered by stationary sources of electricity that are directly connected to the vehicles through wires. Electric traction allows the use of regenerative braking, in which the motors are used as brakes and become generators that transform the motion of, usually, a train into electrical power that is then fed back into the lines. This system is particularly advantageous in mountainous operations, as descending vehicles can produce a large portion of the power required for those ascending. This regenerative system is only viable if the system is large enough to utilise the power generated by descending vehicles.

Battery electric bus by BYD in the Netherlands

In the systems above motion is provided by a rotary electric motor. However, it is possible to "unroll" the motor to drive directly against a special matched track. These linear motors are used in maglev trains which float above the rails supported by magnetic levitation. This allows for almost no rolling resistance of the vehicle and no mechanical wear and tear of the train or track. In addition to the high-performance control systems needed, switching and curving of the tracks becomes difficult with linear motors, which to date has restricted their operations to high-speed point to point services.

## Properties of EVs

### Components

The type of battery, the type of traction motor and the motor controller design vary according to the size, power and proposed application, which can be as small as a motorized shopping cart or wheelchair, through pedelecs, electric motorcycles and scooters, neighborhood electric vehicles, industrial fork-lift trucks and including many hybrid vehicles.

### Energy Sources

Although EVs have few direct emissions, all rely on energy created through electricity generation, and will usually emit pollution and generate waste, unless it is generated by renewable source power plants. Since EVs use whatever electricity is delivered by their electrical utility/grid operator, EVs can be made more or less efficient, polluting and expensive to run, by modifying the electrical generating stations. This would be done by an electrical utility under a government energy policy, in a timescale negotiated between utilities and government.

Fossil fuel vehicle efficiency and pollution standards take years to filter through a nation's fleet of vehicles. New efficiency and pollution standards rely on the purchase of new vehicles, often as the current vehicles already on the road reach their end-of-life. Only a few nations set a retirement age for old vehicles, such as Japan or Singapore, forcing periodic upgrading of all vehicles already on the road.

EVs will take advantage of whatever environmental gains happen when a renewable energy generation station comes online, a fossil-fuel power station is decommissioned or upgraded. Conversely, if government policy or economic conditions shifts generators back to use more polluting fossil fuels and internal combustion engine vehicles (ICEVs), or more inefficient sources, the reverse can happen. Even in such a situation, electrical vehicles are still more efficient than a comparable amount of fossil fuel vehicles. In areas with a deregulated electrical energy market, an electrical vehicle owner can choose whether to run his electrical vehicle off conventional electrical energy sources, or strictly from renewable electrical energy sources (presumably at an additional cost), pushing other consumers onto conventional sources, and switch at any time between the two.

## Issues With Batteries

Old: Banks of conventional lead-acid car batteries are still commonly used for EV propulsion

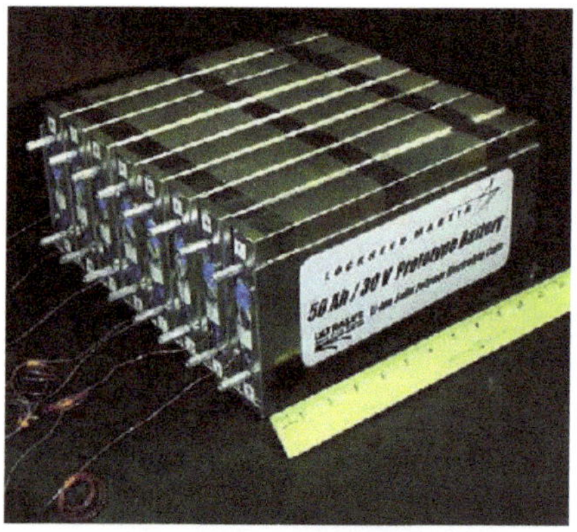

75 watt-hour/kilogram lithium ion polymer battery prototypes. Newer Li-poly cells provide up to 130 Wh/kg and last through thousands of charging cycles.

## Efficiency

Because of the different methods of charging possible, the emissions produced have been quantified in different ways. Plug-in all-electric and hybrid vehicles also have different consumption characteristics.

## Electromagnetic Radiation

Electromagnetic radiation from high performance electrical motors has been claimed to be associated with some human ailments, but such claims are largely unsubstantiated except for extremely high exposures. Electric motors can be shielded within a metallic Faraday cage, but this reduces efficiency by adding weight to the vehicle, while it is not conclusive that all electromagnetic radiation can be contained.

## Charging

### Grid Capacity

If a large proportion of private vehicles were to convert to grid electricity it would increase the demand for generation and transmission, and consequent emissions. However, overall energy consumption and emissions would diminish because of the higher efficiency of EVs over the entire cycle. In the USA it has been estimated there is already nearly sufficient existing power plant and transmission infrastructure, assuming that most charging would occur overnight, using the most efficient off-peak base load sources.

In the UK however, things are different. While National Grid's high-voltage electricity transmission system can currently manage the demand of 1 million electric cars, Steve Holliday (CEO National Grid PLC) said, "penetration up and above that becomes a real issue. Local distribution networks in cities like London may struggle to balance their grids if drivers choose to all plug in their cars at the same time."

### Charging Stations

EVs typically charge from conventional power outlets or dedicated charging stations, a process that typically takes hours, but can be done overnight and often gives a charge that is sufficient for normal everyday usage.

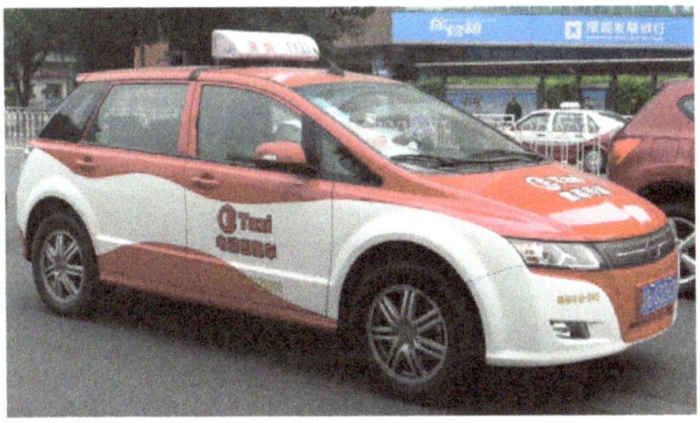

BYD e6 taxi in Shenzhen, China. Recharging in 15 Minutes to 80 Percent

However, with the widespread implementation of electric vehicle networks within large cities, such as those provided by POD Point in the UK and Europe, EV users can plug in their cars whilst at work and leave them to charge throughout the day, extending the possible range of commutes and eliminating range anxiety.

A BYD electric bus in Shanghai at a charging station

A recharging system that avoids the need for a cable is Curb Connect, patented in 2012 by Dr Gordon Dower. In this system, electrical contacts are fitting into curbs, such as angle parking spaces on city streets. When a suitably authorized vehicle is parked so that its front end overhangs the curb, the curb contacts become energized and charging occurs.

A battery electric bus charging station in Geneva, Swiss

Another proposed solution for daily recharging is a standardized inductive charging system such as Evatran's Plugless Power. Benefits are the convenience of parking over the charge station and minimized cabling and connection infrastructure. Qualcomm is trialling such a system in London in early 2012.

Yet another proposed solution for the typically less frequent, long distance travel is "rapid charging", such as the Aerovironment PosiCharge line (up to 250 kW) and the Norvik MinitCharge line (up to 300 kW). Ecotality is a manufacturer of Charging Stations and has partnered with Nissan on several installations. Battery replacement is also proposed as an alternative, although no OEMs including Nissan/Renault have any production vehicle plans. Swapping requires standardization across platforms, models and manufacturers. Swapping also requires many times more battery packs to be in the system.

One type of battery "replacement" proposed, vanadium redox battery, is much simpler: while the latest generation of vanadium redox battery only has an energy density similar to lead-acid, the charge is stored solely in a vanadium-based electrolyte, which can be pumped out and replaced with charged fluid. The vanadium battery system is also a potential candidate for intermediate energy storage in quick charging stations because of its high power density and extremely good endurance in daily use. System cost however, is still prohibitive. As vanadium battery systems are estimated to range between $350–$600 per kWh, a battery that can service one hundred customers in a 24-hour period at 50 kWh per charge would cost $1.8-$3 million.

According to Department of Energy research conducted at Pacific Northwest National Laboratory, 84% of existing vehicles could be switched over to plug-in hybrids without requiring any new grid infrastructure. In terms of transportation, the net result would be a 27% total reduction in emissions of the greenhouse gases carbon dioxide, methane, and nitrous oxide, a 31% total reduction in nitrogen oxides, a slight reduction in nitrous oxide emissions, an increase in particulate matter emissions, the same sulfur dioxide emissions, and the near elimination of carbon monoxide and volatile organic compound emissions (a 98% decrease in carbon monoxide and a 93% decrease in volatile organic compounds). The emissions would be displaced away from street level, where they have "high human-health implications."

## Battery Swapping

Instead of recharging EVs from electric socket, batteries could be mechanically replaced on special stations in a couple of minutes (battery swapping).

Batteries with greatest energy density such as metal-air fuel cells usually cannot be recharged in purely electric way. Instead, some kind of metallurgical process is needed, such as aluminum smelting and similar.

Silicon-air, aluminum-air and other metal-air fuel cells look promising candidates for swap batteries. Any source of energy, renewable or non-renewable, could be used to remake used metal-air fuel cells with relatively high efficiency. Investment in infrastructure will be needed. The cost of such batteries could be an issue, although they could be made with replaceable anodes and electrolyte.

## Chassis Swapping

Instead of replacing batteries, it is possible to replace the entire chassis (including the batteries, electric motor and wheels) of an electric Modular vehicle.

Such a system was patented in 2000 by Dr Gordon Dower and three road-licensed prototypes have been built by the Ridek Corporation in Point Roberts, Washington.

Dr Dower has proposed that an individual might own only the body (or perhaps a few different style bodies) for their vehicle, and would lease the chassis from a pool, thereby reducing the depreciation costs associated with vehicle ownership.

## Other In-development Technologies

Conventional electric double-layer capacitors are being worked to achieve the energy density of lithium ion batteries, offering almost unlimited lifespans and no environmental issues. High-K electric double-layer capacitors, such as EEStor's EESU, could improve lithium ion energy density several times over if they can be produced. Lithium-sulphur batteries offer 250 Wh/kg. Sodium-ion batteries promise 400 Wh/kg with only minimal expansion/contraction during charge/discharge and a very high surface area. Researchers from one of the Ukrainian state universities claim that they have manufactured samples of pseudocapacitor based on Li-ion intercalation process with 318 Wh/kg specific energy, which seem to be at least two times improvement in comparison to typical Li-ion batteries.

## Safety

The United Nations in Geneva (UNECE) has adopted the first international regulation (Regulation 100) on safety of both fully electric and hybrid electric cars, with the intent of ensuring that cars with a high voltage electric power train, such as hybrid and fully EVs, are as safe as combustion-powered cars. The EU and Japan have already indicated that they intend to incorporate the new UNECE Regulation in their respective rules on technical standards for vehicles

There is a growing concern about the safety of EVs, given the demonstrated tendency of the Lithium-ion battery, most promising for EV use because of its high energy density, to overheat, possibly leading to fire or explosion, especially when damaged in a crash. The U.S. National Highway Traffic Safety Administration opened a defect investigation of the Chevy Volt on November 25, 2011 amid concerns over the risk of battery fires in a crash. At that time, automotive consulting firm CNW Marketing Research reported a decline in consumer interest in the Volt, citing the fires as having made an impact on consumer perception. Consumer response impelled GM to make safety enhancements to the battery system in December, and the NHTSA closed its investigation on January 20, 2012, finding the matter satisfactorily resolved with "no discernible defect trend" remaining. The agency also announced it has developed interim guidance to increase awareness and identify appropriate safety measures regarding electric vehicles for the emergency response community, law enforcement officers, tow truck operators, storage facilities and consumers.

## Advantages and Disadvantages of EVs

## Environmental

EVs release no tail pipe air pollutants at the place where they are operated. They also typically generate less noise pollution than an internal combustion engine vehicle, whether at rest or in motion. The energy that electric and hybrid cars consume is usually generated by means that have environmental impacts. Nevertheless, adaption of EVs would have a significant net environmental benefit, except in a few countries that continue to rely on older coal fired power plants for the bulk of their electricity generation throughout the life of the car.

There are special kind of electric vehicles named SAFA TEMPO in Nepal that help lower the pollution created by vehicles. These vehicles are powered by electricity - usually charged batteries - rather than oil or gas and currently heavily promoted by the government to facilitate environmental and vehicle management issues.. Electric motors don't require oxygen, unlike internal combustion engines; this is useful for submarines and for space rovers.

## Mechanical

Electric motors are mechanically very simple and often achieve 90% energy conversion efficiency over the full range of speeds and power output and can be precisely controlled. They can also be combined with regenerative braking systems that have the ability to convert movement energy back into stored electricity. This can be used to reduce the wear on brake systems (and consequent brake pad dust) and reduce the total energy requirement of a trip. Regenerative braking is espccially effective for start-and-stop city use.

Tesla Model S chassis with drive motor

They can be finely controlled and provide high torque from rest, unlike internal combustion engines, and do not need multiple gears to match power curves. This removes the need for gearboxes and torque converters.

EVs provide quiet and smooth operation and consequently have less noise and vibration than internal combustion engines. While this is a desirable attribute, it has also evoked concern that the absence of the usual sounds of an approaching vehicle poses a danger to blind, elderly and very young pedestrians. To mitigate this situation, automakers and individual companies are developing systems that produce warning sounds when EVs are moving slowly, up to a speed when normal motion and rotation (road, suspension, electric motor, etc.) noises become audible.

Cutaway view of a Tesla Model S drive motor

## Energy Resilience

Electricity can be produced from a variety of sources, therefore it gives the greatest degree of energy resilience.

## Energy Efficiency

EV 'tank-to-wheels' efficiency is about a factor of 3 higher than internal combustion engine vehicles. Energy is not consumed while the vehicle is stationary, unlike internal combustion engines which consume fuel while idling. However, looking at the well-to-wheel efficiency of EVs, their total emissions, while still lower, are closer to an efficient gasoline or diesel in most countries where electricity generation relies on fossil fuels.

Well-to-wheel efficiency of an EV has less to do with the vehicle itself and more to do with the method of electricity production. A particular EV would instantly become twice as efficient if electricity production were switched from fossil fuel to a wind or tidal primary source of energy. Thus, when "well-to-wheels" is cited, one should keep in mind that the discussion is no longer about the vehicle, but rather about the entire energy supply infrastructure - in the case of fossil fuels this should also include energy spent on exploration, mining, refining, and distribution.

## Cost of Recharge

According to General Motors, as reported by CNN Money, the GM Volt will cost "less than purchasing a cup of your favorite coffee" to recharge. The Volt should cost less than 2 cents per mile to drive on electricity, compared with 12 cents a mile on gasoline at a price of $3.60 a gallon. This means a trip from Los Angeles to New York would cost $56 on electricity, and $336 with gasoline. This would be the equivalent to paying 60 cents a gallon of gas.

The reality is that the cost of operating an EV varies wildly depending on the part of the world in which the owner lives. In some locations an EV costs less to drive than a comparable gas-powered vehicle, as long as the higher initial purchase-price is not factored in (i.e. a pure comparison of gasoline cost to electricity cost). In the USA, however, in states which have a tiered electricity rate schedule, "fuel" for EVs today costs owners significantly more than fuel for a comparable gas-powered vehicle. A study done by Purdue University found that in California most users already reach the third pricing tier for electricity each month, and adding an EV could push them into the fourth or fifth (highest, most expensive) tier, meaning that they will be paying in excess of $.45 cents per KWH for electricity to recharge their vehicle. At this price, which is higher than the average electricity price in the US, it is dramatically more expensive to drive a pure-EV than it is to drive a traditional pure-gas powered vehicle. "The objective of a tiered pricing system is to discourage consumption. It's meant to get you to think about turning off your lights and conserving electricity. In California, the unintended consequence is that plug-in hybrid cars won't be economical under this system," said Tyner (the author), whose findings were published in the online version of the journal Energy Policy.

## Stabilization of The Grid

Since EVs can be plugged into the electric grid when not in use, there is a potential for battery powered vehicles to even cut the demand for electricity by feeding electricity *into* the grid from their batteries during peak use periods (such as midafternoon air conditioning use) while doing most of their charging at night, when there is unused generating capacity. This vehicle-to-grid (V2G) connection has the potential to reduce the need for new power plants, as long as vehicle owners do not mind reducing the life of their batteries, by being drained by the power company during peak demand.

Furthermore, our current electricity infrastructure may need to cope with increasing shares of variable-output power sources such as windmills and PV solar panels. This variability could be addressed by adjusting the speed at which EV batteries are charged, or possibly even discharged.

Some concepts see battery exchanges and battery charging stations, much like gas/petrol stations today. Clearly these will require enormous storage and charging potentials, which could be manipulated to vary the rate of charging, and to output power during shortage periods, much as diesel generators are used for short periods to stabilize some national grids.

## Range

Many electric designs have limited range, due to the low energy density of batteries compared to the fuel of internal combustion engined vehicles. EVs also often have long recharge times compared to the relatively fast process of refueling a tank. This is further complicated by the current scarcity of public charging stations. "Range anxiety" is a label for consumer concern about EV range.

## Heating of EVs

In cold climates, considerable energy is needed to heat the interior of a vehicle and to defrost the windows. With internal combustion engines, this heat already exists as waste combustion heat diverted from the engine cooling circuit. This process offsets the greenhouse gases' external costs. If this is done with battery EVs, the interior heating requires extra energy from the vehicles' batteries. Although some heat could be harvested from the motor or motors and battery, their greater efficiency means there is not as much waste heat available as from a combustion engine.

However, for vehicles which are connected to the grid, battery EVs can be preheated, or cooled, with little or no need for battery energy, especially for short trips.

Newer designs are focused on using super-insulated cabins which can heat the vehicle using the body heat of the passengers. This is not enough, however, in colder climates as a driver delivers only about 100 W of heating power. A heat pump system, capable of cooling the cabin during summer and heating it during winter, seems to be the most practical and promising way of solving the thermal management of the EV. Ricardo Arboix introduced (2008) a new concept based on the principle of combining the thermal-management of the EV-battery with the thermal-management of the cabin using a heat pump system. This is done by adding a third heat-exchanger, thermally connected with the battery-core, to the traditional heat pump/air conditioning system used in previous EV-models like the GM EV1 and Toyota RAV4 EV. The concept has proven to bring several benefits, such as prolonging the life-span of the battery as well as improving the performance and overall energy-efficiency of the EV.

## Electric Public Transit Efficiency

Shifts from private to public transport (train, trolleybus, personal rapid transit or tram) have the potential for large gains in efficiency in terms of individual miles per kWh.

Research shows people do prefer trams, because they are quieter and more comfortable and perceived as having higher status. Therefore, it may be possible to cut liquid fossil fuel consumption in cities through the use of electric trams. Trams may be the most energy-efficient form of public transportation, with rubber wheeled vehicles using 2/3 more energy than the equivalent tram, and run on electricity rather than fossil fuels.

In terms of net present value, they are also the cheapest—Blackpool trams are still running after 100-years, but combustion buses only last about 15-years.

## Incentives and Promotion

Many governments offer incentives to promote the use of electric vehicles, with the goals of reducing air pollution and oil consumption. Some incentives intend to increase purchases of electric vehicles by offsetting the purchase price with a grant. Other incentives include lower tax rates or exemption from certain taxes, and investment in charging infrastructure.

## Future

Ferdinand Dudenhoeffer, head of the Centre of Automotive Research at the Gelsenkirchen University of Applied Sciences in Germany, said that "by 2025, all passenger cars sold in Europe will be electric or hybrid electric".

Rimac Concept One, electric supercar, since 2013. 0 to 100 km/h in 2.8 seconds, 1088 hp

Tesla Model S, since 2012. 0 to 100 km/h in 3 seconds, recharging in 30 minutes to 80 percent, range 500 km

## Improved Batteries

First, advances in lithium ion batteries, in large part driven by the consumer electronics industry, allow full-sized, highway-capable EVs to be propelled as far on a single charge as conventional cars go on a single tank of gasoline. Lithium batteries have been made safe, can be recharged in minutes instead of hours, and now last longer than the typical vehicle. The production cost of these lighter, higher-capacity lithium batteries is gradually decreasing as the technology matures and production volumes increase.

Toyota Motors Corporation is trying to replace the current lithium ion battery with solid-state battery technology by 2020. The solid-state battery replaces the liquid electrolyte with a solid electrolyte.

Rechargeable lithium-air batteries potentially offer increased range over other types and are a current topic of research.

### Battery Management and Intermediate Storage

Another improvement is to decouple the electric motor from the battery through electronic control, employing supercapacitors to buffer large but short power demands and regenerative braking energy. The development of new cell types combined with intelligent cell management improved both weak points mentioned above. The cell management involves not only monitoring the health of the cells but also a redundant cell configuration (one more cell than needed). With sophisticated switched wiring it is possible to condition one cell while the rest are on duty.

## Slow Sand Filter

Slow sand filters are used in water purification for treating raw water to produce a potable product. They are typically 1 to 2 metres deep, can be rectangular or cylindrical in cross section and are used primarily to treat surface water. The length and breadth of the tanks are determined by the flow rate desired by the filters, which typically have a loading rate of 0.1 to 0.2 metres per hour (or cubic metres per square metre per hour).

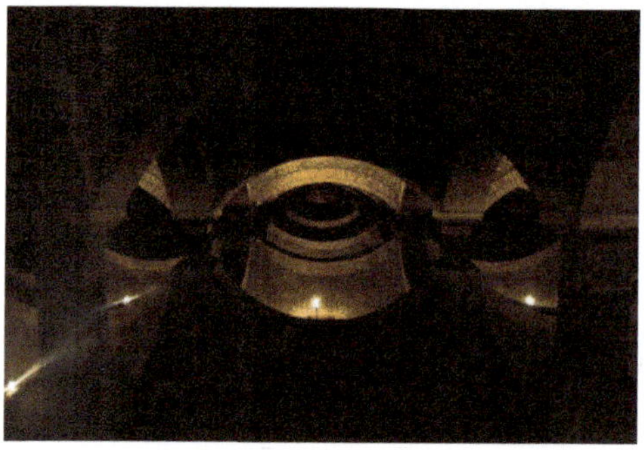

Slow filter in "Filtry Lindleya", Warsaw

A profile of slow sand filter used in a water purification plant.

Slow sand filters differ from all other filters used to treat drinking water in that they work by using a complex biological film that grows naturally on the surface of the sand. The sand itself does not perform any filtration function but simply acts as a substrate, unlike its counterparts for UV and pressurized treatments. Although they are often the preferred technology in many developing countries because of their low energy requirements and robust performance, they are also used to treat water in some developed countries, such as the UK, where they are used to treat water supplied to London. Slow sand filters now are also being tested for pathogen control of nutrient solutions in hydroponic systems.

## History

Original map by John Snow showing the clusters of cholera cases in the London epidemic of 1854.

The first documented use of sand filters to purify the water supply dates to 1804, when the owner of a bleachery in Paisley, Scotland, John Gibb, installed an experimental filter, selling his unwanted surplus to the public. This method was refined in the following two decades by engineers working for private water companies, and it culminated in the first treated public water supply in the world, installed by engineer James Simpson for the Chelsea Waterworks Company in London in 1829. This installation provided filtered water for every resident of the area, and the network design was widely copied throughout the United Kingdom in the ensuing decades.

The practice of water treatment soon became mainstream, and the virtues of the system were made starkly apparent after the investigations of the physician John Snow during the 1854 Broad Street cholera outbreak. Snow was sceptical of the then-dominant miasma theory that stated that diseases were caused by noxious "bad airs". Although the germ theory of disease had not yet been developed, Snow's observations led him to discount the prevailing theory. His 1855 essay *On the Mode of Communication of Cholera* conclusively demonstrated the role of the water supply in spreading the cholera epidemic in Soho, with the use of a dot distribution map and statistical proof to illustrate the connection between the quality of the water source and cholera cases. His data convinced the local council to disable the water pump, which promptly ended the outbreak.

The Metropolis Water Act introduced the regulation of the water supply companies in London, including minimum standards of water quality for the first time. The Act "made provision for securing the supply to the Metropolis of pure and wholesome water", and required that all water be "effectually filtered" from 31 December 1855. This was followed up with legislation for the mandatory inspection of water quality, including comprehensive chemical analyses, in 1858. This legislation set a worldwide precedent for similar state public health interventions across Europe. The Metropolitan Commission of Sewers was formed at the same time, water filtration was adopted throughout the country, and new water intakes on the Thames were established above Teddington Lock.

## Method of Operation

Slow sand filters work through the formation of a gelatinous layer (or biofilm) called the hypogeal layer or *Schmutzdecke* in the top few millimetres of the fine sand layer. The *Schmutzdecke* is formed in the first 10–20 days of operation and consists of bacteria, fungi, protozoa, rotifera and a range of aquatic insect larvae. As an epigeal biofilm ages, more algae tend to develop and larger aquatic organisms may be present including some bryozoa, snails and Annelid worms. The surface biofilm is the layer that provides the effective purification in potable water treatment, the underlying sand providing the support medium for this biological treatment layer. As water passes through the hypogeal layer, particles of foreign matter are trapped in the mucilaginous matrix and soluble organic material is adsorbed. The contaminants are metabolised by the bacteria, fungi and protozoa. The water produced from an exemplary slow sand filter is of excellent quality with 90-99% bacterial cell count reduction.

The raw water is let in to the filter room slowly from the pipe on the right. The water will pass through the sand layers down to the bottom of this room. The Schmutzdecke layer can be observed in this picture.

Slow sand filters slowly lose their performance as the biofilm thickens and thereby reduces the rate of flow through the filter. Eventually, it is necessary to refurbish the filter. Two methods are commonly used to do this. In the first, the top few millimetres of fine sand is scraped off to expose a new layer of clean sand. Water is then decanted back into the filter and re-circulated for a few hours to allow a new biofilm to develop. The filter is then filled to full volume and brought back into service. The second method, sometimes called wet harrowing, involves lowering the water level to just above the hypogeal layer, stirring the sand; thus precipitating any solids held in that layer and allowing the remaining water to wash through the sand. The filter column is then filled to full capacity and brought back into service. Wet harrowing can allow the filter to be brought back into service more quickly.

## Features

Slow sand filters have a number of unique qualities:

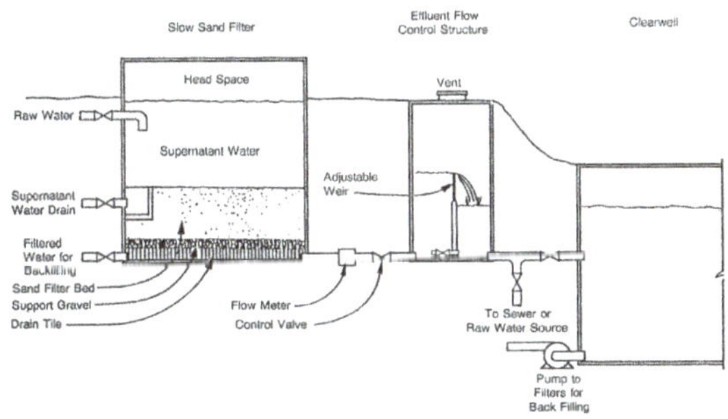


Typical configuration of a housed slow sand filter system

1. Unlike other filtration methods, slow sand filters use biological processes to clean the water, and are non-pressurized systems. Slow sand filters do not require chemicals or electricity to operate.

2. Cleaning is traditionally done by use of a mechanical scraper, which is usually driven into the filter bed once the bed has been dried out. However, some slow sand filter operators use a method called "wet harrowing", where the sand is scraped while still under water, and the water used for cleaning is drained to waste.

3. For municipal systems there usually is a certain degree of redundancy, since it is desirable for the maximum required throughput of water to be achievable with one or more beds out of service.

4. Slow sand filters require relatively low turbidity levels to operate efficiently. In summer conditions and in conditions when the raw water is turbid, blinding of the filters occurs more quickly and pre-treatment is recommended.

5. Unlike other water filtration technologies that produce water on demand, slow sand filters produce water at a slow, constant flow rate and are usually used in conjunction with a storage tank for peak usage. This slow rate is necessary for healthy development of the biological processes in the filter.

Artificial infiltration works on the principles of slow sand filters

While many municipal water treatment works will have 12 or more beds in use at any one time, smaller communities or households may only have one or two filter beds.

In the base of each bed is a series of herringbone drains that are covered with a layer of pebbles which in turn is covered with coarse gravel. Further layers of sand are placed on top followed by a thick layer of fine sand. The whole depth of filter material may be more than 1 metre in depth, the majority of which will be fine sand material. On top of the sand bed sits a supernatant layer of unpurified water.

## Advantages

- As they require little or no mechanical power, chemicals or replaceable parts, and they require minimal operator training and only periodic maintenance, they are often an appropriate technology for poor and isolated areas.
- Slow sand filters, due to their simple design, may be created DIY. DIY-slow sand filters have been used in Afghanistan and other countries to aid the poor.
- Slow sand filters are recognized by the World Health Organization, Oxfam, and the United States Environmental Protection Agency as being superior technology for the treatment of surface water sources. According to the World Health Organization, "Under suitable circumstances, slow sand filtration may be not only the cheapest and simplest but also the most efficient method of water treatment."

## Disadvantages

- Due to the low filtration rate, slow sand filters require extensive land area for a large municipal system. Many municipal systems in the U.S. initially used slow sand filters, but as cities have grown they subsequently installed rapid sand filters, due to increased demand for drinking water.

# Treatment Pond

A treatment pond is a type of pond which is constructed to treat lightly polluted water or wastewater. A treatment pond may be used in combination with a rainwater reservoir to form an ecological, self-purifying irrigation reservoir or swimming pond.

A treatment pond next to an irrigation reservoir, forming a self-purifying reservoir

## Function

Waste treatment ponds are very efficient even though their design has not changed in years. The plants work well in nearly all environments and can treat most forms of wastewater produced. Treatment ponds are used for smaller purposes in the United States but are used for major cities and factories in the rest of the world, as it is a simple and effective method to treat water. Many improvements have been made to improve their effectiveness and efficiency at turning harmful chemicals and sewage into less harmful forms of the previous. With ponds there is a variety of key elements to take care of in order to be properly maintained for the wastewater to be properly treated. Factors such as oxygen levels and flow through the ponds are just a few of the critical factors to properly maintaining the ponds. These factors are constantly monitored to maintain the pond and gain information on what factors are working the best in order to speed up the breakdown of harmful chemicals at the most efficient rate. Treatment ponds that are aerobic, anaerobic or facultative flow through basins are used in order to begin the water treatment process. Anaerobic ponds are used to treat water with industrial waste whereas aerobic ponds are used to treat water contaminated from other plant processes. Facultative ponds are good to treat raw wastewater and water that was not treated properly in the other ponds because it is both aerobic and anaerobic. Aerobic ponds are saturated with dissolved oxygen throughout the entire pond and are shallow enough for light to penetrate the entire depth of the pond. Anaerobic ponds allow the solid waste to settle down at the bottom as a waste sludge while dissolving organic material.

## Design

### Wastewater Characteristics

Typically fresh wastewater has a musty odor and is cloudy with a grayish coloration as it is fresh and untreated. When deprived of oxygen the water becomes black from further with a foul odor. The odor is that of rotten eggs and the water becomes corrosive inviting problems to form in the pipes, so usually treatment workers try to keep it from getting too oxygen deprived while in pipes. Temperature can be anywhere between 45–75 °F (7–24 °C), but the temperature of water entering the system usually does not affect the pond, because it is designed to handle a variety of temperatures. The pH of the system is a huge factor in pond water characteristics with the water either being acidic 0.0–7.0 or basic 7.0–14.0. The pH levels are monitored to make sure that the water does not get too acidic and start to corrode the pipes. It is most desirable for the water to be more basic so as to prevent corrosion from the acid. Another set of characteristics is the contents of the water, like solids, suspended solids and dissolved solids. Dissolved solids are able to pass through most filters whereas suspended solids are just particles suspended in the water column easily caught in strainers or filters. Solids is just a general term for what would remain if all of the water was evaporated from a sample of the wastewater. To get into more technical terms colloidal solids are particles that are so small that they will not settle by gravity on their own. Settleable solids are pretty much self-explanatory as they are the solids that will settle on their own in a given period of time.

### Sizing

When being designed there are several factors that go into planning the build such as climate of the region, cost and the best way to process the waste in the water. Once the pond has been designed the next thing that must be considered is the location of the pond. The size depends on the type of wastewater entering the pond, the type of pond and location in order to begin to size it. Depths usually range from between four and twenty feet deep with the type of bacteria and total amount of waste the plant must treat being the determining factor in the design of depth. With a key factor in determination of size being how much wastewater is going to be treated by the pond the question of how much surface area is needed varies from a few thousand feet to many acres. Anaerobic ponds must be designed to be far deeper than either aerobic or facultative ponds, this is in order to decrease the oxygen levels in the deeper parts of the ponds so the bacteria can do their jobs, eating the waste. Aerobic ponds must be shallower with a great surface area so that more oxygen can dissolve into the water giving the bacteria enough oxygen to properly function. Facultative ponds meet somewhere in the middle in their design with a large surface area and a fairly deep design to allow for both bacteria to function properly.Usually, common reed or Phragmites australis are used in treatment ponds (e.g., in greywater treatment systems to purify wastewater). In self-purifying water reservoirs (used to purify rainwater), other plants are used as well. These reservoirs need to be filled with 1/4 lavastones and water-purifying plants.

## Ecology

Aerobic bacteria grow in the aerobic ponds and are able to decompose organic waste into oxidized products. Cyanobacteria are typically grown in aerobic ponds in order to provide the bacteria in the pond with plenty of oxygen. Anaerobic ponds contain anaerobic bacteria, which are able to breakdown complex organic waste into basic compounds that are less harmful to the environment. There are several kinds of algae that grow in treatment ponds such as green, red and brown algae. In most ponds both bacteria and algae are needed in order to maximize the decomposition of chemicals. Bacteria and algae work in tandem with bacteria breaking down the complex compounds and chemicals. Then the algae decompose the newly formed compounds and are able to produce oxygen for the bacteria that need it. Several types of invertebrates are present in the ponds here they maintain algae levels and help allow the sediment to settle on the bottom. With mosquitos being a problem for those who maintain the ponds the addition of mosquito fish and several other types of predatory fish is good in order to keep the bugs in check. They are primarily used for mosquito control as treatment ponds are fantastic breeding grounds for mosquitos. By adding the fish the ponds bug issues are controlled in a natural and environmentally friendly way without having to use harmful chemicals.

## Other Types

Rooftop water purifying ponds are being used on rooftops. These green roofs can be built from a simple substrate or with plant-based ponds. Plants used include calamus, Menyanthes trifoliata, Mentha aquatica.

Treatment ponds use a wide variety of plants, depending on the local climate and other conditions. Plants are generally chosen which are indigenous, for environmental reasons and optimum performance. In addition to water purifying (de-nutrifying) plants, plants that supply oxygen, and shade are added in ecological water catchments, ponds. This allows a complete ecosystem. Local bacteria and non-predatory fish may be added to eliminate pests. The bacteria are usually grown by submerging straw in water and allowing bacteria from the surrounding air to form on it. Plants are divided in four water depth-zones:

1. A water depth from 0–20 cm. *Iris pseudacorus*, *Sparganium erectum* may be placed here (temperate climates).
2. A water depth from 40–60 cm. Stratiotes aloides, *Hydrocharis morsus-ranae* may be placed here (temperate climates).
3. A water depth from 60–120 cm. *Nymphea alba* may be placed here (temperate climates).
4. A submerged water depth. *Myriophyllum spicatum* may be placed here (temperate climates).

## References


- David B. Sandalow, ed. (2009). Plug-In Electric Vehicles: What Role for Washington? (1st. ed.). The Brookings Institution. pp. 2–5. ISBN 978-0-8157-0305-1.
- Cobb, Jeff (2016-06-22). "Global Plug-in Car Sales Cruise Past 1.5 Million". HybridCars.com. Retrieved 2016-06-22.
- International Council on Clean Transportation (ICCT) (2016). "European Vehicle Market Statistics - Pocketbook 2015/16" (PDF). ICCT. Retrieved 2016-06-17.
- "Worldwide Sales of Toyota Hybrids Surpass 9 Million Units" (Press release). Toyota City, Japan: Toyota. 2016-05-20. Retrieved 2016-05-22.
- Millikin, Mike (2016-05-20). "Worldwide sales of Toyota hybrids surpass 9 million units; Prius family accounts for 63%". Green Car Congress. Retrieved 2016-05-22.
- Sebastian Blanco (2015-11-17). "UCS: Well-to-wheel, EVs cleaner than pretty much all gas cars". Autoblog (website). Retrieved 2015-11-22.
- Nikolaisen, Per-Ivar . "12 mega dams that changed the world (in Norwegian)" In English Teknisk Ukeblad, 17 January 2015. Retrieved 22 January 2015.
- Hunt, Robert (1887). British Mining: A Treatise in the History, Discovery, Practical Development, and Future Prospects of Metalliferous Mines of the United Kingdom (2nd ed.). London: Crosby Lockwood and Co. p. 505. Retrieved 2 May 2015.
- Al-Hassani, Salim. "800 Years Later: In Memory of Al-Jazari, A Genius Mechanical Engineer". Muslim Heritage. The Foundation for Science, Technology, and Civilisation. Retrieved 30 April 2015.
- Jeff Cobb (2015-06-01). "Renault-Nissan And Leaf Lead All In Global EV Proliferation". HybridCars.com. Retrieved 2015-06-14.
- "Ladbare biler i Norge mai, 2015" [Rechargeable cars in Norway in May 2015] (in Norwegian). Grønn bil. June 2015. Retrieved 2015-06-27.
- Dan Rutherford (2014-04-04). "Hybrids break through in the Japan auto market". International Council on Clean Transportation (ICCT). Retrieved 2015-01-25.
- Nealer, Rachael; Reichmuth, David; Anair, Don (November 2015). "Cleaner Cars from Cradle to Grave: How Electric Cars Beat Gasoline Cars on Lifetime Global Warming Emissions" (PDF). Union of Concerned Scientists (UCS). Retrieved 2014-11-22.
- Matthe, Roland; Eberle, Ulrich (2014-01-01). "The Voltec System - Energy Storage and Electric Propulsion". Retrieved 2014-05-04.
- Alan Ohnsman (2014-09-09). "Californians Propel Plug-In Car Sales With 40% of Market". Bloomberg News. Retrieved 2014-09-09.
- Staff (2014-04-02). "Elbilsalget i mars slo alle rekorder" [Electric vehicle sales in March broke all records] (in Norwegian). Grønn bil. Retrieved 2014-04-03.
- Matthew Klippenstein (2014-04-08). "One Percent Of Norway's Cars Are Already Plug-In Electrics". Green Car Reports. Retrieved 2014-04-09.
- Staff (2014-01-08). "Over 20.000 ladbare biler på norske veier" [Over 20,000 rechargeable electric cars on Norwegian road] (in Norwegian). Grønn bil. Retrieved 2014-01-13.

5

# Acts and Laws of Environmental Engineering

The branch of environmental engineering law deals exclusively with complex environmental issues like asbestos baseline survey, soil contamination and assessment and remediation, environmental land use restriction etc. This chapter delineates the local and international environmental laws whose sound knowledge is required to shape environmental engineering technologies. The chapter expounds on the National Environmental Policy Act, Environmental Quality Improvement Act and environmental protocols that have to be followed by environmental engineers.

## National Environmental Policy Act

The National Environmental Policy Act (NEPA) is a United States environmental law that promotes the enhancement of the environment and established the President's Council on Environmental Quality (CEQ). The law was enacted on January 1, 1970. As the bill was an early step towards the development of the United States's environmental policy, NEPA is referred to as the "environmental Magna Carta".

NEPA's most significant outcome was the requirement that all executive federal agencies prepare environmental assessments (EAs) and environmental impact statements (EISs). These reports state the potential environmental effects of proposed federal agency actions. NEPA does not apply to the President, Congress, or the federal courts.

### History

NEPA grew out of the increased appreciation and concern for the environment that resulted from the 1969 Santa Barbara oil spill. During this time, environmental interest group efforts and the movement resulting from Rachel Carson's book, *Silent Spring*, helped to pass the Wilderness, Clean Air, and Clean Water Acts. Another major driver for enacting NEPA were the 1960s freeway revolts, a series of protests that occurred in response to the bulldozing of many communities and ecosystems during the construction of the Interstate Highway System.

Since its passage, NEPA has been applied to any major project, whether on a federal, state, or local level, that involves federal funding, work performed by the federal government, or permits issued by a federal agency. Court decisions have expanded the requirement for NEPA-related environmental studies to include actions where permits issued by a federal agency are required regardless of whether federal funds are spent to implement the action. This legal interpretation is based on the rationale that obtaining a permit from a federal agency requires one or more federal employees (or contractors in some instances) to process and approve a permit application, inherently resulting in federal funds being expended to support the proposed action, even if no federal funds are directly allocated to finance the particular action.

## Contents

The preamble to NEPA reads:

> "To declare national policy which will encourage productive and enjoyable harmony between man and his environment; to promote efforts which will prevent or eliminate damage to the environment and biosphere and stimulate the health and welfare of man; to enrich the understanding of the ecological systems and natural resources important to the Nation; and to establish a Council on Environmental Quality."

NEPA contains three sections: the first section outlines national environmental policies and goals; the second establishes provisions for federal agencies to enforce such policies and goals; and the third establishes of the CEQ in the Executive Office of the President.

The purpose of NEPA is to ensure that environmental factors are weighted equally when compared to other factors in the decision making process undertaken by federal agencies and to establish a national environmental policy. The act also promotes the CEQ to advise the President in the preparation of an annual report on the progress of federal agencies in implementing NEPA. It also established the CEQ to advise the president on environmental policy and the state of the environment.

NEPA establishes this national environmental policy by requiring federal agencies to prepare an environmental impact statement to accompany reports and recommendations for Congressional funding. This impact statement is known as an EIS. NEPA is an action-forcing piece of legislation, meaning the act itself does not carry any criminal or civil sanctions, and therefore, all enforcement of NEPA must occur through the court system. In practice, a project is required to meet NEPA guidelines when a federal agency provides any portion of financing for the project. However, review of a project by a federal employee can be viewed as a federal action, and in such a case, it requires NEPA-compliant analysis performance.

NEPA covers a vast array of federal agency actions, but the act does not apply to state action where there is a complete absence of federal influence or funding. Exemptions and exclusions are also present within NEPA's guidelines, including specific federal projects detailed in legislation and EPA exemptions. Exemptions also apply when compliance with other environmental laws require an impact analysis similar to that mandated by NEPA. Such laws can include but are not limited to the Clean Air Act, Resource Conservation and Recovery Act, Safe Drinking Water Act, and the Federal Insecticide, Fungicide, and Rodenticide Act.

## Process

The NEPA process is the evaluation of the relevant environmental effects of a federal project or action mandated by NEPA. This process begins when an agency develops a proposal addressing a need to take action. If it is determined that the proposed action is covered under NEPA, there are three levels of analysis that a federal agency must undertake to comply with the law. These three levels include the preparation of a Categorical Exclusion (CatEx), an environmental assessment (EA), a Finding of No Significant Impact (FONSI), and, finally, the preparation and drafting of an environmental impact statement (EIS).

## Preparation of a Categorical Exclusion

A Categorical Exclusion (CatEx) is a list of actions an agency has determined do not individually or cumulatively affect the quality of the human environment (40 C.F.R. §1508.4). If a proposed action is included in an agency's CatEx, the agency must make sure that no extraordinary circumstances might cause the proposed action to affect the environment. Extraordinary circumstances include effects on endangered species, protected cultural sites, and wetlands. If the proposed action is not included in the description provided in the CatEx, an EA must be prepared. Actions similar to the proposed one may have been found to be environmentally neutral in previous EAs and their implementation, and so an agency may amend their implementing regulations to include the action as a CatEx. In this case, the drafted agency procedures are published in the *Federal Register* and a public comment period is required. A CatEx for one agency cannot be used by another agency unless this procedure has been followed.

## Preparation of an Environmental Assessment and a Finding of No Significant Impact

EAs are concise public documents that include the need for a proposal, a list of alternatives, and a list of agencies and persons consulted in the proposal's drafting. The purpose of an EA is to determine the significance of the proposal's environmental outcomes and to look at alternatives of achieving the agency's objectives. An EA is supposed to provide sufficient evidence and analysis for determining whether to prepare an EIS, aid an agency's compliance with NEPA when no EIS is necessary, and it facilitates preparing an EIS when one is necessary.

Most agency procedures do not require public involvement prior to finalizing an EA document; however, agencies advise that a public comment period is considered at the draft EA stage. EAs need to be of sufficient length to ensure that the underlying decision to prepare an EIS is legitimate, but they should not attempt to substitute an EIS.

If no substantial effects on the environment are found after investigation and the drafting of an EA, the agency must produce a FONSI. This document explains why an action will not have a significant effect on the human environment and includes the EA or a summary of the EA that supports the FONSI determination.

## Preparation of an Environmental Impact Statement

If it is determined that a proposed federal action does not fall within a designated CatEx or does not qualify for a FONSI, then the responsible agency must prepare an EIS. The purpose of an EIS is to help public officials make informed decisions based on the relevant environmental consequences and the alternatives available. The drafting of an EIS includes public party, outside party, and other federal agency input concerning its preparation. These groups subsequently comment on the draft EIS.

An EIS is required to describe the environmental impacts of the proposed action, any adverse environmental impacts that cannot be avoided should the proposal be implemented, the reasonable alternatives to the proposed action, the relationship between local short-term uses of man's environment along with the maintenance and enhancement of long-term productivity, and any irreversible and irretrievable commitments of resources that would be involved in the proposed action.

An agency may undertake the drafting of an EIS without the initial drafting of the EA. This may happen if the agency believes that the action will have a significant impact on the human or natural environment or if the action is considered an environmentally controversial issue.

## Council on Environmental Quality

The Council on Environmental Quality (CEQ) was modeled after the Council of Economic Advisers, a group of advisers to the President, created by the Employment Act of 1946. Shortly after NEPA was signed into law, President Richard Nixon expanded the CEQ's mandate by executive order. He directed the CEQ to issue guidelines for the proper preparation of an EIS and to assemble and coordinate federal programs related to environmental quality. The Council was placed within the Executive Office of the President of the United States and is composed of three President-appointed members, which are subsequently confirmed by the Senate.

The CEQ has played a key part in the development of the EIS process. Its initial guidelines were issued in 1971, and required each federal department and agency to adopt its own guidelines consistent those established by CEQ. These guidelines did not carry the status of formal agency regulations, but were often held in the court of law as such. Eventually President Jimmy Carter authorized an executive order to adopt regulations rather than simple guidelines on EIS preparation. However, the CEQ had no authority to enforce its regulations.

The CEQ regulations begin by calling for agencies to integrate NEPA requirements with other planning requirements as soon as possible. This ensures that all decisions are reflective of environmental values, avoids potential delays, and eliminates potential future conflicts. NEPA's action-forcing provision, Section 102(2)(C), stipulates that an EIS shall be "included in every recommendation or report on proposals for legislation and other major Federal actions significantly affecting the quality of the human environment, a detailed statement by the responsible official..."

The CEQ has taken measures within the past several years to prepare advisory documentation explaining the general structure of NEPA and the nature of cumulative impacts, among other advisories. The CEQ maintains a website at NEPA.

### Recent Developments

In January 2015, Republican Senator Rob Portman, introduced the "Federal Permitting Improvement Act of 2015". The bill would establish several requirements and procedures designed to expedite the completion of NEPA-related reviews, including the creation of a "Federal Permitting Improvement Council".

## Environmental Quality Improvement Act

The Environmental Quality Improvement Act of 1970 is a United States environmental law which was passed to work in conjunction with the National Environmental Policy Act of 1969 (NEPA). One of the two major purposes of the Act was to authorize the creation of an Office of Environmental Quality to provide the professional and administrative staff needed for the Council on Environmental Quality. The second major purpose was to "assure that each Federal department and agency conducting or supporting public works activities which affect the environment shall implement the policies under existing law". To accomplish these purposes, the act gave more responsibilities to the Chairman of the Council on Environmental Quality in his new role as Director of the Office of Environmental Quality.

### Legislative History

The Environmental Quality Improvement Act of 1970 started out as H.R. 4148 in the U.S. House of Representatives Public Works Committee in April 1970. It was taken up by the U.S. Senate Public Works Committee in October 1969 and went to a joint conference between the houses in March 1969. The bill passed the Senate with a vote of 80 to 0 on March 24 and the House on March 25 with a vote of 358 to 0. The lack of any nay votes shows there was broad support for the passage of this bill.

The bill also had six amendments from the years 1973 to 1984 but the amendments mostly dealt with providing additional funding in various years over the amount specified by NEPA. The amounts ranged from a low of an additional $44,000.00/ year for

1982-1984 to a high of $3,000,000/year additional funds for 1980 and 1981. Besides providing additional funds, Pub. L. 98-581 (10-30-1984) amended the Environmental Quality Improvement Act to create a fund called the Office of Environmental Quality Management Fund (42 US Code 4375(a)). The fund was established to finance study contracts that are sponsored by both the OEQ and at least one other Federal agency. The fund can also fund Federal Interagency environmental projects such as a task force that the OEQ is participating in. The fund is allowed to receive advance payments from other agencies to finance the proposed study, project or activity.

## Major Provisions

In Section 4372 (a), (b), and (c) of the act, the Office of Environmental Quality (OEQ) is established as an office within the Executive Office of the President of the United States with the Chairman of the Council on Environmental Quality (CEQ) as the Director of the OEQ. In addition, the President of the United States also picks a Deputy Director for the OEQ who is also to be approved by the Senate. While the Director is allowed to employ the amount of people he may need to carry out the work of the OEQ, he is limited to hiring only ten specialists or experts. These specialists or experts also cannot be paid more than a GS-18 pay scale. The function of the OEQ is to provide support for the CEQ and to help the Director carry out his new duties as given in Section 4372 (d). The OEQ and the CEQ are simply known jointly as the Council on Environmental Quality and there is not a separate listing for the OEQ within the Office of the President or a separate website from the CEQ.

Another major area in the Environmental Quality Improvement Act was the creation of six new duties or functions given to the director of OEQ. Besides the director fulfilling the duties to the President of the United States and the CEQ under NEPA, the director had to make sure that the following provisions of U.S. Code 4372 (d) 2-7 were carried out as well:.

4372 (d ) 2: assisting the Federal agencies and departments in appraising the effectiveness of existing and proposed facilities, programs, policies, and activities of the Federal Government, and those specific major projects designated by the President which do not require individual project authorization by Congress, which affect environmental quality;

4372 (d) 3: reviewing the adequacy of existing systems for monitoring and predicting environmental changes in order to achieve effective coverage and efficient use of research facilities and other resources;

4372 (d) 4: promoting the advancement of scientific knowledge of the effects of actions and technology on the environment and encourage the development of the means to prevent or reduce adverse effects that endanger the health and well-being of man;

4372 (d) 5: assisting in coordinating among the Federal departments and agencies those programs and activities which affect, protect, and improve environmental quality;

4372 (d) 6: assisting the Federal departments and agencies in the development and interrelationship of environmental quality criteria and standards established through the Federal Government;

and 4372 (d) 7: collecting, collating, analyzing, and interpreting data and information on environmental quality, ecological research, and evaluation.

The last major provision is that the annual Environmental Quality Report required by NEPA (Public Law 91-190) will be sent to each standing committee that has jurisdiction over any part that might be contained in the report. The Environmental Quality Report, as described in NEPA, relates the status and condition of air, aquatic, and terrestrial environments, future and current trends in regards to those environments, how adequate those resources are for providing our economic needs, a review of significant activities all levels of governments, other entities, and individuals are doing in the area of conservation, effect on the environment, and in developing natural resources, and also how existing programs that are lacking will be improved along with recommendations for legislation. The preparation of this report would be an example of one of the duties from NEPA that the Director of the OEQ would be responsible for as well.

## Impact

As part of the duties given to the Director by this act, the staff of the OEQ works on many guidances and notices that assist federal agencies and entities with implementing policy on environmental issues. Some recent examples are:

Final Guidance on Improving the Process for Preparing Efficient and Timely Environmental Reviews Under the National Environmental Policy Act - the guidance goes over crucial techniques that agencies are not fully employing in preparing environmental reviews. These environmental reviews (environmental impact assessments and environmental assessments) are important in understanding the environmental impact and what can be done to mitigate it.

Final Guidance, "Federal Greenhouse Gas Accounting and Reporting" - This is the final notice to agencies to help them comply with an annual acoounting for Greenhouse Gases as required by Executive Order #13514. The purpose of the annual accounting is to gauge over a number of years if an agency's reduction plan for greenhouse gases is effective. It also notes there will be revisions to this guidance as better techniques are created.

Instructions for Implementing Sustainable Locations for Federal Facilities in Accordance With Executive Order 13514 - This gives agencies directions on how to incorporate principles of sustainable locations into the decision making process for where a building should be built or leased. A building that fits in with future regional transportation plans, is near public transport, and is in a pedestrian friendly area would have a more sustainable location.

As noted in Section 4374 (d)5, one of the duties of the Director and OEQ is assisting with other agencies in activities that will help improve, protect, and affect environmental quality. An example of this is the Chair of the Council on Environmental Quality, Nancy Sutley (the Director of OEQ as well), chaired an Interagency Ocean Policy Task Force that among other things looked into effective coastal and marine spatial planning. Another employee of the OEQ also served on the steering committee of this taskforce to help coordinate its activities.

# Environmental Law

Environmental law - or "environmental and natural resources law" - is a collective term describing the network of treaties, statutes, regulations, and common and customary laws addressing the effects of human activity on the natural environment.

## Regulatory Subjects

The broad category of "environmental law" may be broken down into a number of more specific regulatory subjects. While there is no single agreed-upon taxonomy, the core environmental law regimes address environmental pollution. A related but distinct set of regulatory regimes, now strongly influenced by environmental legal principles, focus on the management of specific natural resources, such as forests, minerals, or fisheries. Other areas, such as environmental impact assessment, may not fit neatly into either category, but are nonetheless important components of environmental law.

## Impact Assessment

Environmental impact assessment (EA) is the term used for the assessment of the environmental consequences (positive and negative) of a plan, policy, program, or concrete projects prior to the decision to move forward with the proposed action. In this context, the term 'environmental impact assessment' (EIA) is usually used when applied to concrete projects by individuals or companies and the term 'strategic environmental assessment' (SEA) applies to policies, plans and programmes most often proposed by organs of state (Fischer, 2016). Environmental assessments may be governed by rules of administrative procedure regarding public participation and documentation of decision making, and may be subject to judicial review.

## Air Quality

Air quality laws govern the emission of air pollutants into the atmosphere. A specialized subset of air quality laws regulate the quality of air inside buildings. Air quality laws are often designed specifically to protect human health by limiting or eliminating airborne pollutant concentrations. Other initiatives are designed to address broader ecological problems, such as limitations on chemicals that affect the ozone layer, and emissions trading programs to address acid rain or climate change. Regulatory efforts include identifying and categorizing air pollutants, setting limits on acceptable emissions levels, and dictating necessary or appropriate mitigation technologies.

Industrial air pollution now regulated by air quality law.

## Water Quality

Water quality laws govern the release of pollutants into water resources, including surface water, ground water, and stored drinking water. Some water quality laws, such as drinking water regulations, may be designed solely with reference to human health. Many others, including restrictions on the alteration of the chemical, physical, radiological, and biological characteristics of water resources, may also reflect efforts to protect aquatic ecosystems more broadly. Regulatory efforts may include identifying and categorizing water pollutants, dictating acceptable pollutant concentrations in water resources, and limiting pollutant discharges from effluent sources. Regulatory areas include sewage treatment and disposal, industrial and agricultural waste water management, and control of surface runoff from construction sites and urban environments.

A typical stormwater outfall, subject to water quality law.

## Waste Management

Waste management laws govern the transport, treatment, storage, and disposal of all manner of waste, including municipal solid waste, hazardous waste, and nuclear waste, among many other types. Waste laws are generally designed to minimize or eliminate the uncontrolled dispersal of waste materials into the environment in a manner that may cause ecological or biological harm, and include laws designed to reduce the generation of waste and promote or mandate waste recycling. Regulatory efforts include identifying and categorizing waste types and mandating transport, treatment, storage, and disposal practices.

A municipal landfill, operated pursuant to waste management law.

## Contaminant Cleanup

Environmental cleanup laws govern the removal of pollution or contaminants from environmental media such as soil, sediment, surface water, or ground water. Unlike pollution control laws, cleanup laws are designed to respond after-the-fact to environmental contamination, and consequently must often define not only the necessary response actions, but also the parties who may be responsible for undertaking (or paying for) such actions. Regulatory requirements may include rules for emergency response, liability allocation, site assessment, remedial investigation, feasibility studies, remedial action, post-remedial monitoring, and site reuse.

Oil spill emergency response, governed by environmental cleanup law.

## Chemical Safety

Chemical safety laws govern the use of chemicals in human activities, particularly man-made chemicals in modern industrial applications. As contrasted with media-oriented environmental laws (e.g., air or water quality laws), chemical control laws seek to manage the (potential) pollutants themselves. Regulatory efforts include banning specific chemical constituents in consumer products (e.g., Bisphenol A in plastic bottles), and regulating pesticides.

## Water Resources

Water resources laws govern the ownership and use of water resources, including surface water and ground water. Regulatory areas may include water conservation, use restrictions, and ownership regimes.

An irrigation ditch, operated in accordance with water resources law.

## Mineral Resources

Mineral resource laws cover several basic topics, including the ownership of the mineral resource and who can work them. Mining is also affected by various regulations regarding the health and safety of miners, as well as the environmental impact of mining.

## Forest Resources

A timber operation, regulated by forestry law.

Forestry laws govern activities in designated forest lands, most commonly with respect to forest management and timber harvesting. Ancillary laws may regulate forest land acquisition and prescribed burn practices. Forest management laws generally adopt management policies, such as multiple use and sustained yield, by which public forest resources are to be managed. Governmental agencies are generally responsible for planning and implementing forestry laws on public forest lands, and may be involved in forest inventory, planning, and conservation, and oversight of timber sales. Broader initiatives may seek to slow or reverse deforestation.

## Wildlife and Plants

Wildlife laws govern the potential impact of human activity on wild animals, whether directly on individuals or populations, or indirectly via habitat degradation. Similar laws may operate to protect plant species. Such laws may be enacted entirely to protect biodiversity, or as a means for protecting species deemed important for other reasons. Regulatory efforts may including the creation of special conservation statuses, prohibitions on killing, harming, or disturbing protected species, efforts to induce and support species recovery, establishment of wildlife refuges to support conservation, and prohibitions on trafficking in species or animal parts to combat poaching.

## Fish and Game

Fish and game laws regulate the right to pursue and take or kill certain kinds of fish and wild animal (game). Such laws may restrict the days to harvest fish or game, the number of animals caught per person, the species harvested, or the weapons or fishing gear used. Such laws may seek to balance dueling needs for preservation and harvest and to manage both environment and populations of fish and game. Game laws can provide a legal structure to collect license fees and other money which is used to fund conservation efforts as well as to obtain harvest information used in wildlife management practice.

## Important Principles

Environmental law has developed in response to emerging awareness of and concern over issues impacting the entire world. While laws have developed piecemeal and for a variety of reasons, some effort has gone into identifying key concepts and guiding principles common to environmental law as a whole. The principles discussed below are not an exhaustive list and are not universally recognized or accepted. Nonetheless, they represent important principles for the understanding of environmental law around the world.

## Sustainable Development

Defined by the United Nations Environment Programme as "development that meets the needs of the present without compromising the ability of future generations to meet their own needs," sustainable development may be considered together with the concepts of "integration" (development cannot be considered in isolation from sustainability) and "interdependence" (social and economic development, and environmental protection, are interdependent). Laws mandating environmental impact assessment and requiring or encouraging development to minimize environmental impacts may be assessed against this principle.

The modern concept of sustainable development was a topic of discussion at the 1972 United Nations Conference on the Human Environment (Stockholm Conference), and the driving force behind the 1983 World Commission on Environment and Development (WCED, or Bruntland Commission). In 1992, the first UN Earth Summit resulted in the Rio Declaration, Principle 3 of which reads: "The right to development must be fulfilled so as to equitably meet developmental and environmental needs of present and future generations." Sustainable development has been a core concept of international environmental discussion ever since, including at the World Summit on Sustainable Development (Earth Summit 2002), and the United Nations Conference on Sustainable Development (Earth Summit 2012, or Rio+20).

## Equity

Defined by UNEP to include intergenerational equity - "the right of future generations to enjoy a fair level of the common patrimony" - and intragenerational equity - "the right of all people within the current generation to fair access to the current generation's entitlement to the Earth's natural resources" - environmental equity considers the present generation under an obligation to account for long-term impacts of activities, and to act to sustain the global environment and resource base for future generations. Pollution control and resource management laws may be assessed against this principle.

## Transboundary Responsibility

Defined in the international law context as an obligation to protect one's own environment, and to prevent damage to neighboring environments, UNEP considers transboundary responsibility at the international level as a potential limitation on the rights of the sovereign state. Laws that act to limit externalities imposed upon human health and the environment may be assessed against this principle.

## Public Participation and Transparency

Identified as essential conditions for "accountable governments . . ., industrial concerns," and organizations generally, public participation and transparency are presented by UNEP as requiring "effective protection of the human right to hold and express opinions and to seek, receive and impart ideas," "a right of access to appropriate, comprehensible and timely information held by governments and industrial concerns on economic and social policies regarding the sustainable use of natural resources and the protection of the environment, without imposing undue financial burdens upon the applicants and with adequate protection of privacy and business confidentiality," and "effective judicial and administrative proceedings." These principles are present in environmental impact assessment, laws requiring publication and access to relevant environmental data, and administrative procedure.

## Precautionary Principle

One of the most commonly encountered and controversial principles of environmental law, the Rio Declaration formulated the precautionary principle as follows:

> In order to protect the environment, the precautionary approach shall be widely applied by States according to their capabilities. Where there are threats of serious or irreversible damage, lack of full scientific certainty shall not be used as a reason for postponing cost-effective measures to prevent environmental degradation.

The principle may play a role in any debate over the need for environmental regulation.

## Prevention

> The concept of prevention . . . can perhaps better be considered an overarching aim that gives rise to a multitude of legal mechanisms, including prior assessment of environmental harm, licensing or authorization that set out the conditions for operation and the consequences for violation of the conditions, as well as the adoption of strategies and policies. Emission limits and other product or process standards, the use of best available techniques and similar techniques can all be seen as applications of the concept of prevention.

## Polluter Pays Principle

The polluter pays principle stands for the idea that "the environmental costs of economic activities, including the cost of preventing potential harm, should be internalized rather than imposed upon society at large." All issues related to responsibility for cost for environmental remediation and compliance with pollution control regulations involve this principle.

## History

Early examples of legal enactments designed to consciously preserve the environment, for its own sake or human enjoyment, are found throughout history. In the common law, the primary protection was found in the law of nuisance, but this only allowed for private actions for damages or injunctions if there was harm to land. Thus smells emanating from pig stys, strict liability against dumping rubbish, or damage from exploding dams. Private enforcement, however, was limited and found to be woefully inadequate to deal with major environmental threats, particularly threats to common resources. During the "Great Stink" of 1858, the dumping of sewerage into the River Thames began to smell so ghastly in the summer heat that Parliament had to be evacuated. Ironically, the Metropolitan Commission of Sewers Act 1848 had allowed the Metropolitan Commission for Sewers to close cesspits around the city in an attempt to "clean up" but this simply led people to pollute the river. In 19 days, Parliament passed a further Act to build the London sewerage system. London also suffered from terrible air pollution, and this culminated in the "Great Smog" of 1952, which in turn triggered its on legislative response: the Clean Air Act 1956. The basic regulatory structure was to set limits on emissions for households and business (particularly burning coal) while an inspectorate would enforce compliance.

Notwithstanding early analogues, the concept of "environmental law" as a separate and distinct body of law is a twentieth-century development. The recognition that the natural environment was fragile and in need of special legal protections, the translation of that recognition into legal structures, the development of those structures into a larger body of "environmental law," and the strong influence of environmental law on natural resource laws, did not occur until about the 1960s. At that time, numerous influences - including a growing awareness of the unity and fragility of the biosphere; increased public concern over the impact of industrial activity on natural resources and human health; the increasing strength of the regulatory state; and more broadly the advent and success of environmentalism as a political movement - coalesced to produce a huge new body of law in a relatively short period of time. While the modern history of environmental law is one of continuing controversy, by the end of the twentieth century environmental law had been established as a component of the legal landscape in all developed nations of the world, many developing ones, and the larger project of international law.

## Controversy

Environmental law is a continuing source of controversy. Debates over the necessity, fairness, and cost of environmental regulation are ongoing, as well as regarding the appropriateness of regulations vs. market solutions to achieve even agreed-upon ends.

Allegations of scientific uncertainty fuel the ongoing debate over greenhouse gas regulation, and are a major factor in debates over whether to ban particular pesticides. In cases where the science is well-settled, it is not unusual to find that corporations intentionally hide or distort the facts, or sow confusion.

It is very common for regulated industry to argue against environmental regulation on the basis of cost. Difficulties arise in performing cost-benefit analysis of environmental issues. It is difficult to quantify the value of an environmental value such as a healthy ecosystem, clean air, or species diversity. Many environmentalists' response to pitting economy vs. ecology is summed up by former Senator and founder of Earth Day Gaylord Nelson, "The economy is a wholly owned subsidiary of the environment, not the other way around." Furthermore, environmental issues are seen by many as having an ethical or moral dimension, which would transcend financial cost. Even so, there are some efforts underway to systemically recognize environmental costs and assets, and account for them properly in economic terms.

While affected industries spark controversy in fighting regulation, there are also many environmentalists and public interest groups who believe that current regulations are inadequate, and advocate for stronger protection. Environmental law conferences - such as the annual Public Interest Environmental Law Conference in Eugene, Oregon - typically have this focus, also connecting environmental law with class, race, and other issues.

An additional debate is to what extent environmental laws are fair to all regulated parties. For instance, researchers Preston Teeter and Jorgen Sandberg highlight how smaller organizations can often incur disproportionately larger costs as a result of environmental regulations, which can ultimately create an additional barrier to entry for new firms, thus stifling competition and innovation.

## Around The World

## International Law

Global and regional environmental issues are increasingly the subject of international law. Debates over environmental concerns implicate core principles of international law and have been the subject of numerous international agreements and declarations.

Customary international law is an important source of international environmental law. These are the norms and rules that countries follow as a matter of custom and they are so prevalent that they bind all states in the world. When a principle becomes customary law is not clear cut and many arguments are put forward by states not wishing to be bound. Examples of customary international law relevant to the environment include the duty to warn other states promptly about icons of an environmental nature and environmental damages to which another state or states may be exposed, and Principle 21 of the Stockholm Declaration ('good neighbourliness' or sic utere).

Numerous legally binding international agreements encompass a wide variety of issue-areas, from terrestrial, marine and atmospheric pollution through to wildlife and biodiversity protection. International environmental agreements are generally multilateral (or sometimes bilateral) treaties (a.k.a. convention, agreement, protocol, etc.). Protocols are subsidiary agreements built from a primary treaty. They exist in many areas of international law but are especially useful in the environmental field, where they may be used to regularly incorporate recent scientific knowledge. They also permit countries to reach agreement on a framework that would be contentious if every detail were to be agreed upon in advance. The most widely known protocol in international environmental law is the Kyoto Protocol, which followed from the United Nations Framework Convention on Climate Change.

While the bodies that proposed, argued, agreed upon and ultimately adopted existing international agreements vary according to each agreement, certain conferences, including 1972's United Nations Conference on the Human Environment, 1983's World Commission on Environment and Development, 1992's United Nations Conference on Environment and Development and 2002's World Summit on Sustainable Development have been particularly important. Multilateral environmental agreements sometimes create an International Organization, Institution or Body responsible for implementing the agreement. Major examples are the Convention on International Trade in Endangered Species of Wild Fauna and Flora (CITES) and the International Union for Conservation of Nature (IUCN).

International environmental law also includes the opinions of international courts and tribunals. While there are few and they have limited authority, the decisions carry much weight with legal commentators and are quite influential on the development of international environmental law. One of the biggest challenges in international decisions is to determine an adequate compensation for environmental damages. The courts include the International Court of Justice (ICJ); the international Tribunal for the Law of the Sea (ITLOS); the European Court of Justice; European Court of Human Rights and other regional treaty tribunals.

## Africa

According to the International Network for Environmental Compliance and Enforcement (INECE), the major environmental issues in Africa are "drought and flooding, air pollution, deforestation, loss of biodiversity, freshwater availability, degradation of soil and vegetation, and widespread poverty." The U.S. Environmental Protection Agency (EPA) is focused on the "growing urban and industrial pollution, water quality, electronic waste and indoor air from cookstoves." They hope to provide enough aid on concerns regarding pollution before their impacts contaminate the African environment as well as the global environment. By doing so, they intend to "protect human health, particularly vulnerable populations such as children and the poor." In order to accomplish these goals in Africa, EPA programs are focused on strengthening the ability to enforce environmental laws as well as public compliance to them. Other programs work on developing stronger environmental laws, regulations, and standards.

## Asia

The Asian Environmental Compliance and Enforcement Network (AECEN) is an agreement between 16 Asian countries dedicated to improving cooperation with environmental laws in Asia. These countries include Cambodia, China, Indonesia, India, Maldives, Japan, Korea, Malaysia, Nepal, Philippines, Pakistan, Singapore, Sri Lanka, Thailand, Vietnam, and Lao PDR.

## European Union

The European Union issues secondary legislation on environmental issues that are valid throughout the EU (so called regulations) and many directives that must be implemented into national legislation from the 28 member states (national states). Examples are the Regulation (EC) No. 338/97 on the implementation of CITES; or the Natura 2000 network the centerpiece for nature & biodiversity policy, encompassing the bird Directive (79/409/EEC/ changed to 2009/147/EC)and the habitats directive (92/43/EEC). Which are made up of multiple SACs (Special Areas of Conservation, linked to the habitats directive) & SPAs (Special Protected Areas, linked to the bird directive), throughout Europe.

EU legislation is ruled in Article 249 Treaty for the Functioning of the European Union (TFEU). Topics for common EU legislation are:

- Climate change
- Air pollution
- Water protection and management
- Waste management
- Soil protection
- Protection of nature, species and biodiversity
- Noise pollution
- Cooperation for the environment with third countries (other than EU member states)
- Civil protection

## Middle East

The U.S. Environmental Protection Agency is working with countries in the Middle East to improve "environmental governance, water pollution and water security, clean fuels and vehicles, public participation, and pollution prevention."

## Oceania

The main concerns on environmental issues in the Oceanic Region are "illegal releases of air and water pollutants, illegal logging/timber trade, illegal shipment of hazardous wastes, including e-waste and ships slated for destruction, and insufficient institutional structure/lack of enforcement capacity". The Secretariat of the Pacific Regional Environmental Programme (SPREP) is an international organization between Australia, the Cook Islands, FMS, Fiji, France, Kiribati, Marshall Islands, Nauru, New Zealand,

Niue, Palau, PNG, Samoa, Solomon Island, Tonga, Tuvalu, USA, and Vanuatu. The SPREP was established in order to provide assistance in improving and protecting the environment as well as assure sustainable development for future generations.

## Australia

The Environment Protection and Biodiversity Conservation Act 1999 is the center piece of environmental legislation in the Australian Government. It sets up the "legal framework to protect and manage nationally and internationally important flora, fauna, ecological communities and heritage places". It also focuses on protecting world heritage properties, national heritage properties, wetlands of international importance, nationally threatened species and ecological communities, migratory species, Commonwealth marine areas, Great Barrier Reef Marine Park, and the environment surrounding nuclear activities. *Commonwealth v Tasmania* (1983), also known as the "Tasmanian Dam Case", is the most influential case for Australian environmental law.

## Brazil

The Brazilian government created the Ministry of Environment in 1992 in order to develop better strategies of protecting the environment, use natural resources sustainably, and enforce public environmental policies. The Ministry of Environment has authority over policies involving environment, water resources, preservation, and environmental programs involving the Amazon.

## Canada

The Department of the Environment Act establishes the Department of the Environment in the Canadian government as well as the position Minister of the Environment. Their duties include "the preservation and enhancement of the quality of the natural environment, including water, air and soil quality; renewable resources, including migratory birds and other non-domestic flora and fauna; water; meteorology;" The Environmental Protection Act is the main piece of Canadian environmental legislation that was put into place March 31, 2000. The Act focuses on "respecting pollution prevention and the protection of the environment and human health in order to contribute to sustainable development." Other principle federal statutes include the Canadian Environmental Assessment Act, and the Species at Risk Act. When provincial and federal legislation are in conflict federal legislation takes precedence, that being said individual provinces can have their own legislation such as Ontario's Environmental Bill of Rights, and Clean Water Act.

## China

According to the U.S. Environmental Protection Agency, "China has been working with great determination in recent years to develop, implement, and enforce a solid environmental law framework. Chinese officials face critical challenges in effectively implementing the laws, clarifying the roles of their national and provincial governments, and strengthening the operation of their legal system." Explosive economic and industrial growth in China has led to significant environmental degradation, and China is currently in the process of developing more stringent legal controls. The harmonization of Chinese society and the natural environment is billed as a rising policy priority.

## Ecuador

With the enactment of the 2008 Constitution, Ecuador became the first country in the world to codify the Rights of Nature. The Constitution, specifically Articles 10 and 71-74, recognizes the inalienable rights of ecosystems to exist and flourish, gives people the authority to petition on the behalf of ecosystems, and requires the government to remedy violations of these rights. The rights approach is a break away from traditional environmental regulatory systems, which regard nature as property and legalize and manage degradation of the environment rather than prevent it.

The Rights of Nature articles in Ecuador's constitution are part of a reaction to a combination of political, economic, and social phenomena. Ecuador's abusive past with the oil industry, most famously the class-action litigation against Chevron, and the failure of an extraction-based economy and neoliberal reforms to bring economic prosperity to the region has resulted in the election of a New Leftist regime, led by President Rafael Correa, and sparked a demand for new approaches to development. In conjunction with this need, the principle of "Buen Vivir," or good living—focused on social, environmental and spiritual wealth versus material wealth—gained popularity among citizens and was incorporated into the new constitution.

The influence of indigenous groups, from whom the concept of "Buen Vivir" originates, in the forming of the constitutional ideals also facilitated the incorporation of the Rights of Nature as a basic tenet of their culture and conceptualization of "Buen Vivir."

## Egypt

The Environmental Protection Law outlines the responsibilities of the Egyptian government to "preparation of draft legislation and decrees pertinent to environmental management, collection of data both nationally and internationally on the state of the environment, preparation of periodical reports and studies on the state of the environment, formulation of the national plan and its projects, preparation of environmental profiles for new and urban areas, and setting of standards to be used in planning for their development, and preparation of an annual report on the state of the environment to be prepared to the President."

## India

In India, Environmental law is governed by the Environment Protection Act, 1986. This act is enforced by the Central Pollution Control Board and the numerous State Pollution Control Boards. Apart from this, there are also individual legislations specifically enacted for the protection of Water, Air, Wildlife, etc. Such legislations include :-

- The Water (Prevention and Control of Pollution) Act, 1974
- The Water (Prevention and Control of Pollution) Cess Act, 1977
- The Forest (Conservation) Act, 1980
- The Air (Prevention and Control of Pollution) Act, 1981
- Air (Prevention and Control of Pollution) (Union Territories) Rules, 1983
- The Biological Diversity Act, 2002 and the Wild Life Protection Act, 1972.
- Batteries (Management and Handling) Rules, 2001
- Recycled Plastics, Plastics Manufacture and Usage Rules, 1999
- The National Green Tribunal established under the National Green Tribunal Act of 2010 has jurisdiction over all environmental cases dealing with a substantial environmental question and acts covered under the Water (Prevention and Control of Pollution) Act, 1974;
- Water (Prevention and Control of Pollution) Cess Rules, 1978
- Ganga Action Plan, 1986
- The Forest (Conservation) Act, 1980
- The Public Liability Insurance Act, 1991 and the Biological Diversity Act, 2002. The acts covered under Indian Wild Life Protection Act 1972 do not fall within the jurisdiction of the National Green Tribunal. Appeals can be filed in the Hon'ble Supreme Court of India.
- Basel Convention on Control of TransboundaryMovements on Hazardous Wastes and Their Disposal, 1989 and Its Protocols
- Hazardous Wastes (Management and Handling) Amendment Rules, 2003

## Japan

The Basic Environmental Law is the basic structure of Japan's environmental policies replacing the Basic Law for Environmental Pollution Control and the Nature Conservation Law. The updated law aims to address "global environmental problems, urban pollution by everyday life, loss of accessible natural environment in urban areas and degrading environmental protection capacity in forests and farmlands."

The three basic environmental principles that the Basic Environmental Law follows are "the blessings of the environment should be enjoyed by the present generation and succeeded to the future generations, a sustainable society should be created where environmental loads by human activities are minimized, and Japan should contribute actively to global environmental conservation through international cooperation." From these principles, the Japanese government have established policies such as "environmental consideration in policy formulation, establishment of the Basic Environment Plan which describes the directions of long-term environmental policy, environmental impact assessment for development projects, economic measures to encourage activities for reducing environmental load, improvement of social infrastructure such as sewerage system, transport facilities etc., promotion of environmental activities by corporations, citizens and NGOs, environmental education, and provision of information, promotion of science and technology."

## New Zealand

The Ministry for the Environment and Office of the Parliamentary Commissioner for the Environment were established by the Environment Act 1986. These positions are responsible for advising the Minister on all areas of environmental legislation. A common theme of New Zealand's environmental legislation is sustainably managing natural and physical resources, fisheries, and forests. The Resource Management Act 1991 is the main piece of environmental legislation that outlines the government's strategy to managing the "environment, including air, water soil, biodiversity, the coastal environment, noise, subdivision, and land use planning in general."

## Russia

The Ministry of Natural Resources and Environment of the Russian Federation makes regulation regarding "conservation of natural resources, including the subsoil, water bodies, forests located in designated conservation areas, fauna and their habitat, in the field of hunting, hydrometeorology and related areas, environmental monitoring and pollution control, including radiation monitoring and control, and functions of public environmental policy making and implementation and statutory regulation."

## South Africa

## United States

## Vietnam

Vietnam is currently working with the U.S. Environmental Protection Agency on dioxin remediation and technical assistance in order to lower methane emissions. In March 2002, the U.S and Vietnam signed the U.S.-Vietnam Memorandum of Understanding on Research on Human Health and the Environmental Effects of Agent Orange/Dioxin.

## Environmental Protocol

An environmental protocol is a type of international law, "an intergovernmental document intended as legally binding with a primary stated purpose of preventing or managing human impacts on natural resources."

International environmental protocols came to feature in environmental governance after trans-boundary environmental problems became widely perceived in the 1960s. Following the Stockholm Intergovernmental Conference in 1972, creation of international environmental agreements proliferated.

The world's existing political systems pose barriers to the creation of environmental protocols. First, maintenance of sovereignty means that no country can be forced to participate, only urged to do so. Consequently, as French states, "International law has the force of moral suasion, but few real teeth." Second, North-South conflict can block cooperation: the countries in the global South generally see the countries of the North as needing to take responsibility for environmental degradation and make significant changes in their way of living, neither of which the North deems reasonable.

Finally, countries may lack motivation to change their environmental policies due to conflict with other interests, especially economic prosperity. If environmental protocols will cause economic difficulties or harm to a country, it may shirk the protocols while other countries adhere to them, creating a classic free-rider problem. Additionally, environmental protocols may be criticized for scientific uncertainty, or at least a lack of synthesis of scientific information, which may be used for "blocking interests and doing mischief."

Due to these barriers, environmental protocols become an obvious target for several criticisms, such as being slow to produce the desired effects (due to the convention-protocol-ratification-implementation process), tending to the lowest common denominator, and lacking monitoring and enforcement. They can also be criticized for taking an incremental approach where sustainable development principles suggest that environmental concern should be mainstreamed.

Protocols can take flexible approaches to improve effectiveness. One example is the use of sanctions: under the Montreal Protocol, signatories were forbidden to purchase chlorofluorocarbons from nonsignatories, in order to prevent any windfall benefits. Funding has also been used to overcome North-South conflict: members of the Montreal Protocol created a fund of $240 million to redistribute the costs of transition. Differential obligations as seen in the Kyoto Protocol can also encourage wider participation.

While protocols appear to be the ultimate top-down mode of governance, having "scant provisions for public participation," it is widely thought that the influence of transnational networks has been growing Public opinion is relevant, as concern must exist to prompt action and dedication of government resources. Non-governmental organizations also fulfill certain roles, from gathering information and devising policies to mobilising support. Science plays an important part, although Susskind asserts that sometimes this role is diminished by uncertainty, disagreement, and the rise of "adversary science." The business community can also be involved with positive outcomes.

How we view the effectiveness of protocols depends on what we expect from them. With little administrative force or actual power, protocols succeed in increasing government concern, enhancing the contractual environment, and heightening capacity through transfer of assets. Yet as long as sovereignty is intact, environmental protocols will not affect changes in the face of state or public apathy, guarantee national action, or materialize overnight. The progress of international environmental law might be, as Wiener suggests, like the tortoise, slow but steady.

## Environmental Engineering Law

Environmental engineering law is a profession that requires an expertise in both environmental engineering and law. This field includes professionals with both a legal and environmental engineering education. This dual educational requirement is typically satisfied through an ABET accredited degree in environmental engineering and an ABA accredited law degree. Likewise, this profession requires both licensure in professional environmental engineering and admittance to one bar.

Environmental engineering law is the professional application of law, science and engineering principles to improve the environment (air, water, and/or land resources), to provide healthy water, air, and land for human habitation and for other organisms, and to remediate polluted sites. Environmental engineering lawyers seek to promote the advancement of technical engineering knowledge in the legal profession and to enhance informed legal analysis of complex environmental matters.

### Practice Areas

Environmental engineering law professionals offer a sound knowledge base in the fields of both environmental engineering and law to address complex environmental problems which demand both professional technical practice and legal expertise. Areas of practice are continually expanding, but frequently include complex land transactions, such as:

- Brownfields redevelopment
- Asbestos baseline survey and building revaluation due to forthcoming asbestos abatements
- Soil contamination assessment & remediation, the development of a remedial action workplan (RAWP) and engineering controls, including an environmental land use restriction (ELUR)

- Total maximum daily load (TMDL) nutrient loading studies (ex. for NPDES wastewater discharges) and regulatory negotiation of nutrients discharge limits from waste treatment plants, such as phosphorus and nitrogen.

## References


- Eccleston, Charles H. (2008). NEPA and Environmental Planning: Tools, Techniques, and Approaches for Practitioners. CRC Press. ISBN 9780849375590.
- Sullivan, Thomas F. P. (2007). Environmental Law Handbook. Blue Ridge Summit, PA: Government Institutes/Scarecrow Press. ISBN 978-0-86587-024-6.
- Nelson, Gaylord (November 2002). Beyond Earth Day: Fulfilling the Promise. Wisconsin Press. ISBN 0-299-18040-9.
- Hardman Reis, T., Compensation for Environmental Damages Under International Law, Kluwer Law International, The Hague, 2011, ISBN 978-90-411-3437-0.
- Taylor, Prue; Stroud, Lucy; Peteru, Clark (2013). Multilateral Environmental Agreement Negotiator's Handbook: Pacific Region 2013 (PDF). Samoa / New Zealand: Secretariat of the Pacific Regional Environment Programme / New Zealand Centre for Environmental Law, University of Auckland. ISBN 978 982-04-0475-5.
- Surendra Malik, Sudeep Malik. Supreme Court on Environment Law (2015 ed.). India: EBC. ISBN 9789351451914.
- "Environmental Quality Improvement Act of 1970, Pub. L. 91-224, sec. 4371(c) (1), April 3, 1970". Retrieved 12 April 2012.
- "Part 1518-Office of Environmental Quality Management Fund, Federal Register, October 4, 2002." (PDF). Retrieved 30 April 2012.
- "Final Guidance on Improving the Process for Preparing Efficient and Timely Environmental Reviews Under the National Environmental Policy Act, Federal Register. March 3rd, 2012.". Retrieved 25 April 2012.
- "Instructions for Implementing Sustainable Locations for Federal Facilities in Accordance With Executive Order 13514, Federal Register, November 3rd, 2011.". Retrieved 27 April 2012.

# Permissions

All chapters in this book are published with permission under the Creative Commons Attribution Share Alike License or equivalent. Every chapter published in this book has been scrutinized by our experts. Their significance has been extensively debated. The topics covered herein carry significant information for a comprehensive understanding. They may even be implemented as practical applications or may be referred to as a beginning point for further studies.

We would like to thank the editorial team for lending their expertise to make the book truly unique. They have played a crucial role in the development of this book. Without their invaluable contributions this book wouldn't have been possible. They have made vital efforts to compile up to date information on the varied aspects of this subject to make this book a valuable addition to the collection of many professionals and students.

This book was conceptualized with the vision of imparting up-to-date and integrated information in this field. To ensure the same, a matchless editorial board was set up. Every individual on the board went through rigorous rounds of assessment to prove their worth. After which they invested a large part of their time researching and compiling the most relevant data for our readers.

The editorial board has been involved in producing this book since its inception. They have spent rigorous hours researching and exploring the diverse topics which have resulted in the successful publishing of this book. They have passed on their knowledge of decades through this book. To expedite this challenging task, the publisher supported the team at every step. A small team of assistant editors was also appointed to further simplify the editing procedure and attain best results for the readers.

Apart from the editorial board, the designing team has also invested a significant amount of their time in understanding the subject and creating the most relevant covers. They scrutinized every image to scout for the most suitable representation of the subject and create an appropriate cover for the book.

The publishing team has been an ardent support to the editorial, designing and production team. Their endless efforts to recruit the best for this project, has resulted in the accomplishment of this book. They are a veteran in the field of academics and their pool of knowledge is as vast as their experience in printing. Their expertise and guidance has proved useful at every step. Their uncompromising quality standards have made this book an exceptional effort. Their encouragement from time to time has been an inspiration for everyone.

The publisher and the editorial board hope that this book will prove to be a valuable piece of knowledge for students, practitioners and scholars across the globe.

# Index